17.58

WESTERN ENERGY POLICY

WESTERN ENERGY POLICY

the case for competition

Douglas Evans

ST. MARTIN'S PRESS NEW YORK

First published in the United States of America in 1979

ISBN 0-312-86392-6

Library of Congress Cataloging in Publication Data

Evans, Douglas.
  Western energy policy.

  Bibliography: p.
  Includes index.
  1. Energy policy. 2. Petroleum industry and
trade. I. Title.
HD9502.A2E93 1979   333.7   78-23315
ISBN 0-312-86392-6

Contents

List of Tables

Acknowledgements

The genesis of this book lies in a report on British energy policy commissioned by the Economic Research Council, London, in 1976. From that original project grew an awareness of the significance of British policy for Western Europe as a whole and, in turn, Western Europe's dependence on the world energy market. All this was a short step to realising that the energy policies of each of West Germany, Britain and the United States were crucial, not only to those countries' future prosperity but to that of the Western industrial system as a whole. Since my personal interest has lain in the field of commercial policy for some time, there was a certain inevitability about this progression. Nevertheless, it is to the Economic Research Council, and the loan they made available to undertake a report on British energy policy, that this book owes its beginnings, and not least to the persistence of its director, Edward Holloway.

For the conversion of the raw typescript into a book I am indebted to my agent, Andrew Best of Curtis Brown, and to T. M. Farmiloe of the Macmillan Press, who, not for the first time, have willingly bent themselves to the task of producing a topical book in short time on my behalf. For nursing this particular book through the production process so conscientiously I wish to thank Allan Aslett.

In terms of the argument of the book I also wish to acknowledge my debt most especially in the compilation of the British chapter to Professor Kenneth Dam of the University of Chicago and Professor Colin Robinson of the University of Surrey, whose respective books on *Oil Resources* and *The Energy Crisis and British Coal* greatly reinforced my previous convictions. For his early guidance and subsequent suggested amendments I wish to thank J. M. Jefferson, chief economist, Shell International Petroleum Co. Ltd., whose expert and up-to-date knowledge was invaluable. In the last analysis, however, nobody but the author is responsible for the views expressed. It is my hope that this book will provide both the evidence and the principal arguments in a form which will widen the discussion on the issues and ingredients of energy policy and convey something of its pervasive impact on all our futures.

Douglas Evans
London, 1978

Introduction: The General Prospect for World Energy

After the crisis in the world energy market and the world economy generally in 1973–4, with its multiple and continuing ramifications for industrialised and non-industrialised countries alike, public awareness of the gravity of the situation has receded. Yet in 1977 the former US Secretary of State, Dr Henry Kissinger, could warn without any hint of exaggeration that failure to solve international energy problems could bring about the destruction of the current world order. Aside from the more obvious and immediate possibilities that at any time the huge cash surplus of the oil-producing countries is capable of being deployed to disrupt the world monetary and financial system, or that a deepening recession among the industrial countries at least partially derived from the energy crisis may be aggravated by a renewal of national autarky, there are more particular reasons outlined in this introduction for examining the prospects for world energy.

However, the raison d'etre for this book is not so much to rehearse possible scenarios for the future world energy market, which must remain a speculative exercise, but more practically, perhaps, to examine the relative performance of some of the major industrial countries in the sphere of energy policy and to suggest the sort of framework best suited to maximise the general benefit. While enabling readers to examine and judge for themselves the past performance and future prospects for the nations under review, this book makes no secret of its belief in the virtues of a free enterprise framework.

For convenience, the prospects for world energy have been considered first to 1985 and then from 1985 to 2000. Since the first aspect is daily reported and analysed in the newspapers and is encompassed within the main body of the book, it will receive only the briefest treatment here, allowing us to take a more detailed look at the longer range forecasts in the final chapter. All such assessments of global demand and supply are not amenable to precise forecasting — a fact which makes them no less necessary in anticipating the range of possibilities with which we will have to contend.

OUTLOOK TO 1985

The world energy market will continue by and large to be dominated by oil during the next few years and unless the national energy conservation programmes prove more successful than hitherto, the likelihood is that world demand for oil will reach the limit of productive capacity by the late 1980s or early 1990s and by the 1980s potential demand will have very substantially outstripped capacity causing prices to rise fairly dramatically. The crux of the supply problem is likely to arise because of OPEC's inability and unwillingness to maintain the escalation in oil for export and the Soviet Union's switch at some point from a net exporter to net importer of oil. However, because of the major increase in oil production from both the North Sea and Alaska, the approaching bottleneck in supply may not be generally apparent. In the meantime the advent of oil from the North Sea and Alaska will have the effect of stabilising OPEC prices probably right up until 1980. However, this stabilising of prices is conditional on Saudi Arabia wishing to hold down the oil price against the wishes of most of the other members of OPEC. For the present this suits the interests of the Saudi Arabians as well as the consumer nations but the policy of Saudi-led restraint on the part of OPEC could be ended virtually any time if political circumstances changed radically.

In the period roughly encompassed between 1979 and 1985 the combination of an expected increase in world demand for oil and a failure to expand production in the major consuming countries, i.e. the United States and the Soviet Union, is likely to result in an increased reliance on OPEC oil. Even if the present Saudi expansion plans were largely implemented, their excess productive capacity could be exhausted by 1985, marking the end of their ability to act as a price moderator in OPEC. The danger looming up is that the alternative oil supplies look very unlikely to come on stream rapidly enough to alter the situation substantially. At the same time as the limits to expanding OPEC production will be coming through, in about 1980, the growth in North Sea supplies will be beginning to slow down; Alaskan output will have stabilised and the Soviet Union will probably have ceased to be a net exporter of oil to the West. Indeed, though viewed by many as somewhat alarmist, the US Central Intelligence Agency calculates that by 1985 the Soviet Union and Eastern Europe will be net importers of somewhere between 3.5 and 4.5 million barrels per day.

Due to the very long lead times involved and the pattern of delays in nuclear plant construction, the possibility of nuclear energy effectively bridging this gap in oil supplies does not appear very promising. Moreover, natural gas supplies outside OPEC will probably not increase overall by 1985, for despite increased output of natural gas from the North Sea, production is likely to decline in both the United States and Canada. Conversely, coal production is expected to expand in the United States but probably remain static if it does not go into decline in most other Western industrial countries. Among Western industrial nations the only important increase in oil production will take place in Western Europe, that is, production derived from the North Sea.

During this period the non-OPEC, less developed countries will require increasing amounts of oil imports in order to sustain even a moderate rate of economic growth as they are currently structured, possibly up to 4 million barrels a day by 1985. If these requirements are not able to be met, or they prove to be cruelly prohibitive financially, the consequences in human suffering and political and social instability could become very serious indeed for the entire international comity of nations. Among non-OPEC, less developed countries, those likely to register the greatest increases in domestic oil production are Mexico and Egypt, with smaller increases expected from Brazil, Tunisia, Oman, Syria, India and Burma. What may be declared with a fair degree of certainty is that the rising pressure of oil demand on capacity in the early-1980s is bound to cause oil prices to rise well in advance of any actual shortage. On present trends, sizeable price increases appear inevitable from the early-1980s.

There are other, less pessimistic views available and it is illuminating to examine some of the differences which exist between the forecasts of the US Central Intelligence Agency, the US International Trade Commission and the Organisation for Economic Co-operation and Development. The most obvious example is the ITC's belief that there will be no energy supply shortage by 1985 in contradistinction to the CIA study which substantially influenced the original Carter proposals on energy.

The reason behind the ITC's greater degree of optimism over future oil supplies is that they calculate that Saudi Arabia will continue to increase its oil output to keep pace with rising demand, something we can examine more carefully later in this introduction. Moreover, the ITC believes the Soviet Union will be able to remain self-sufficient in oil. The latter conflicts with two CIA reports which suggest that the Soviet Union will have increasing oil production difficulties, more especially an inability to supply other East European countries with adequate supplies by 1985. Since the CIA is generally assumed to have the greater expertise in knowledge of the Soviet bloc countries, it is the CIA analysis which must probably be given greater weight in any assessment on the latter aspect. The CIA fears that mounting demands from East European nations for Middle East oil will sharply affect the pattern of world supply and demand. The ITC counters that the Soviet Union will raise output by new recovery methods and by drilling offshore. This writer is deeply sceptical that the Soviet Union possesses the technology to match the challenge.

Where the ITC may be right in its assessment is in believing that the Soviet Union's outer continental shelf may be the only remaining area where vast reserves of oil and gas may be found in a volume equal to that in the Middle East. To a very large extent its conclusions are based upon likely conservation efforts in the major oil importing countries and upon rising oil production in the North Sea, Alaska and non-Arab nations such as Mexico. The ITC expects the 1985 level of oil output by OPEC not to be very different from the 1976 production figure of 30.4 million barrels per day, though Shell calculates the potential 1985 OPEC export volume significantly higher at 37/38 million barrels per day. The ITC also expects that, both because of Alaskan production and conservation efforts, the United States oil imports in 1985 will total about 7.2 million barrels daily, compared with the present level of more than 9 million barrels and the CIA prediction of 16 million barrels (Shell equivalent figure of 12/13 mbd).

But if there are differences between the CIA and ITC reports, there are possibly even more substantial differences between the CIA and OECD expectations over world oil supplies. Despite these differences, both sources underscore the truth that the world's economies are still inextricably linked with OPEC for the indefinite future. While the United States may be currently preoccupied with attempts to curb consumption through energy taxes, both these reports remind the reader that the marginal supply cost of energy will continue to be established by the economic rent demand made by OPEC.

The differences in estimates between the CIA and OECD studies — the former entitled International Energy Situation Outlook to 1985 and the latter, World Energy Outlook — are as significant for the source of their differences as much as their actual predictions. Where the CIA sees demand for OPEC oil rising to around 50 million barrels a day by 1985 from a base of 31 million barrels a day in 1976, the OECD projects OPEC production at less than 40 million barrels a day by 1985. This crucial discrepancy derives from very different appraisals of the supply-demand situation in two distinct and clearly contrasting areas: the less developed countries and the Communist bloc. Thus in the key sector of the oil import requirements of the Western industrialised nations, whose energy policies this book attempts to analyse, both reports forecast oil imports running at 35 million barrels a day in 1985, up from approximately 26 million barrels a day in 1976. Very significantly, the OECD points out that its predictions rest heavily on a balanced US energy policy combining conservation with supply incentives in the shape of the elimination of price controls on oil and natural gas producers. Since the latter has not yet been adopted by the US Administration, the prediction of both reports may yet prove unfounded with widespread ramifications for the remaining Western industrial nations, as this book cumulatively underscores.

The principal reason why the Carter energy planners, at least initially, seemed to underestimate the need for domestic supply incentives is that a number of them hold to the belief that there is insufficient oil and gas in the United States to justify the effort and the expense. Unconsciously it may also be difficult to reconcile an emphasis on conservation, agreed by everyone to be essential, with the provision of real incentives for present production of domestic oil and gas. The declining real cash flow from the existing reserves in the Lower-48 and south Alaska may even tend to support such a pessimistic view, though it is not the one generally held in the oil industry as a whole.

Unlike the CIA, which does not make policy recommendations, the OECD makes a number of recommendations, especially on US policy, which is so pervasive. The OECD Secretariat, for instance, foresees the possibility that the OECD countries, that is the Western industrialised countries, could reduce oil imports by more than 10 million barrels a day below trend by 1985. The OECD believes that the United States alone could achieve a reduction of its oil import demand that would be equivalent to half of the savings achievable by all member countries, but only by a combination of demand management and accelerated development of domestic supplies. Specifically the OECD advocates for the United States:

> ... elimination of price controls on crude oil and natural gas; accelerated granting of exploration leases and production licenses; moderation of environment requirements for new energy development, particularly coal and nuclear power; incentive measures for application of secondary and tertiary oil recovery techniques, and active promotion of coal utilization by removal of demand constraints.

The OECD was not content with making general policy suggestions but went on to suggest detailed ideas for the United States to achieve import savings by the mid-1980s.

TABLE 1.   *US Potential Import Savings by 1985*
(million barrels a day)

CONSERVATION

Sector	Potential Saving	Policy Proposals
Transportation	0.7	Establishment of 27.5 m.p.g. standard; improved airline load factor.
Household	0.4	Income tax credit for insulation; higher building thermal standards; appliance labelling.
Industrial	0.4	Greater investment tax credits for new equipment; industrial conservation monitoring.
	1.5	

DEVELOPMENT

Oil-Beaufort Sea	0.4	Granting leases; permit for looping of Alaska.
Oil-OCS	1.0	Acceleration of leasing.
Oil-onshore	0.9	Government research and application subsidy to advanced recovery techniques.
Natural Gas & NGL	1.5	Complete decontrol of natural gas prices and acceleration of OCS licensing.
	3.8	

The significance of these figures in the light of the Carter proposals first put to Congress is that they indicate first, the OECD's belief that more than 70 per cent of the potential import savings would derive from increased development and second, that President Carter might be said to have adopted all the proposals on conservation and none of those on development. If the real objective is to reduce energy imports, there is much evidence to suggest that permitting increased prices to flow back into additional exploration and development rather than providing various deterrents to demand is the more effective means to that end.

Among the most foreboding paragraphs in the OECD report where it relates to US energy policy is the following:

> If controls were extended, and especially if the extension is expected by investors early-on, this would serve as a significant disincentive for exploration and development. If essentially the same type of controls were to last the length of the EPCA, or through 1982, the potential level of production could drop by at least 500,000 barrels a day. Obviously the longer the controls remain in effect and the wider the disparity between domestic and world prices, the more deleterious the effect will be upon production and the greater the dependence on imports.

While the CIA and OECD studies do not record striking differences in aggregate figures for demand and supply for 1985, there are important regional variations. Among the most important is the CIA's implied 1985 imported energy figure for the United States which is nearly 4 million barrels a day higher than the OECD's. A 13.4 million barrels a day level of US imports predicted by the CIA for 1985 is more than double the Carter target for the same year.

In calculating the precise security of supply it can be safely predicted that not all of US energy imports in 1985 will be obtained in the form of OPEC oil. However, if liquefied natural gas figures prominently, as the analysis included in the US chapter of this book suggests it might, then practically all US imports of liquefied natural gas up to 1985 will come from OPEC countries. What is significant in the context of the present debate on US energy policy is that the OPEC price of LNG is likely to be at least twice the level the Carter Administration, at least initially, wished to pay domestic gas producers. Deregulation of gas could alter that possibility drastically. Meanwhile it must be remembered that even at the most conservative estimates — those of the OECD — the Western industrial countries will still be importing oil at the rate of 30 million barrels a day in 1980 and 35 million by 1985. This is without taking into account either gas or coal imports, of which the latter may have to be derived from the Soviet bloc to some considerable extent and the former from the politically uncertain North African suppliers of LNG, Algeria and Libya.

This brings us to the source of the principal differences between the OECD and CIA studies, namely their widely different assessment of the supply and demand situation in the non-OPEC, less developed countries and also in the Sino-Soviet blocs. In the case of the CIA estimate of the non-OPEC, less developed countries there can be room for doubt that their oil demand in 1985 will in fact reach twice the level predicted by the OECD for the simple reason that they are realistically not able to afford it. On the question of the Communist bloc countries this author is much more inclined to favour the CIA than the OECD forecasts, given the greater specialist expertise and the semi-disguised nature of the Soviet and Chinese statistics in particular.

The OECD predicts that the Soviet Union and her East Europe satellites will move from a small export position of around 500,000 barrels a day in 1980 to a level essentially in balance by 1985; by contrast the CIA expects the Soviet Union and Eastern Europe to be competing with the West for at least 3.5 million barrels of oil a day by 1985, due to an expected sharp decline in production rather than any appreciable growth in demand. The OECD believes oil exports from China will be at around 500,000 barrels a day in 1980, rising to 1.2 million barrels a day by 1985. By contrast the CIA believes that Chinese oil exports will be negligible by 1985.

It is the combination of unfavourable expectations, that is, a significant growth in its oil import requirements on the part of the Soviet Union and a failure of the Chinese to sustain their recently developed oil export capacity, that explains the CIA's pessimistic view of the world energy situation for the next decade. Meanwhile, if we were to add the total disparity in the projections of the OECD and CIA for the Communist bloc and the non-OPEC, less developed countries, we discover a truly critical figure of 10 million barrels a day between them. It is only by getting behind the reasons for those differences that the respective forecasts can be assessed.

What both the OECD and CIA reports underline, what must be an inescapable conclusion for all energy policy planners in the Western world, is that, whatever hopes may be entertained for alternatives, whatever efforts may be launched to realise those alternatives, the brute fact is that OPEC and Saudi Arabia in particular are likely to remain the residual source of energy for the foreseeable future, probably for most of the balance of this century. When national energy planners attempt to bypass this reality, usually by suggesting that consumers need not pay the price that OPEC imposes (often well below the replacement cost level), they are entering the world of fantasy. The CIA places this approach of OPEC reality as likely to occur around 1983. The crucial paragraph runs:

> Between 1979 and 1985, increasing world demand and stagnating oil production in the major consuming countries will result in increased reliance on OPEC oil. By 1985 we estimate that demand for OPEC oil will reach 47 to 51 million barrels a day. Even if all other OPEC states produce at capacity, Saudi Arabia will be required to produce between 19 and 23 million barrels a day if demand is to be met. This is well above present Saudi capacity of 10 to 11 million barrels a day, and projected 1985 capacity of at most 18 million barrels a day. With the present expansion plans of the Saudis, their excess productive capacity will be exhausted by 1983, and with it their ability to act as price moderator in OPEC.

While both OECD and CIA reports are agreed that some time in the 1980s OPEC prices are likely to rise significantly in real terms, eventually reaching whatever level is required to reflect the long-term cost of the energy systems of the next century, there is a contrary point of view gaining ground. It runs more or less as follows. Ever since OPEC's cartel-like control of world oil prices led to the higher prices of 1973 onwards, many commentators and professional observers have supported the general thesis that the new oil price system would prove unsupportable over the long term. In many cases this view was based on the most impeccable classical economic theory as well as a great body of historical evidence drawn from the history of cartels that defined them as a union of sellers formed to raise prices, a circumstance that historically rarely lasted for very long. The general argument states that both in theory and in practice human greed must eventually work to break up any rigged market. Unfortunately, the logic of the argument breaks down at its base since, as we shall see later, OPEC fails to meet the strict definition of either a cartel or a monopoly. There were other and more particular arguments deployed.

They included the argument that OPEC's political unity was too fragile to stand up to the problems of oil surpluses in the market and that weaker members of OPEC would be bound in time to yield to temptation in the shape of under-the-counter discounting just as soon as they failed to sell their full production. Another argument is that OPEC members, or more accurately, a majority of them, were going into debt so rapidly that their demands for development capital would prove insatiable and must lead in time to competitive selling of oil, not because of any absolute necessity but because of the unmodifiable expectations of their people, at once socially sophisticated but economically naïve. A further argument put forward was that the largest producers in OPEC would be seduced in time, probably sooner rather than later, with counterpart industrial countries with complementary requirements of secure oil supplies to match the supplying nation's need for a stable demand situation. Thus, went these several arguments, the tenuous unity of OPEC would be stripped bare and gradually cracked.

But these predictions so long promulgated have yet to be fulfilled. On the one hand OPEC has managed to sustain a very successful pricing performance, while on the other, and not so fully recognised, the sheer cumulative effect of the major industrial powers, accommodating themselves by various measures to higher world oil prices, have cemented the fabric of the OPEC alliance. In the United States the specific policy measures which have so contributed include higher taxes on crude oil, increases in natural gas prices, 'frontier' pricing incentives for North Slope (Alaska) crude, tax encouragement for investment in conservation, and not least, higher prices for Canadian and North Sea oil.

By late 1977, four years after the Yom Kippur war had precipitated the most severe economic slowdown since the 1930s, there were ominous signs that a second recessionary phase was beginning. A host of commentators soon began to predict a rapid collapse of world energy prices. Their arguments grew essentially out of the proposition that a combination of decreasing demand for oil worldwide and growing productive capacity (notably in Alaska, the North Sea and southern Mexico) will lead by turn to an oil surplus, competitive discounting among OPEC members, a price break in world crude oil and as a final consequence in the chain of causation, significant price reductions in all energy sources.

The foregoing scenario would be a remarkable turnabout to say the least. At a stroke some of the most pressing problems in the world economic system would evaporate. They include the problem of an overall economic slowdown, the payments owing by the less developed countries to OPEC, the threat of the world banking system collapsing, political fears of increasing terrorism in Western Europe financed by oil money and fuelled by inflation and rising unemployment, and the most extreme possible worldwide economic collapse in the early 1980s, all most conveniently banished. Of course there would be losers also of which the oil companies who were over-extended in high cost hydrocarbons, such as oil and gas in the North Sea and the North Slope, would be the most obvious example. Britain, above any other country would be the hardest hit since she had in comparative terms staked the most in the North Sea.

While it is a brave man who does not hedge his bets on such a central question the preceding scenario of a collapse of the OPEC oil price level as described is probably misconceived and ultimately unfounded. In spite of the fact that in *real* terms the price of OPEC oil has fallen — by about $4 per barrel — since the peak, the overall situation is likely to be considerably more favourable to OPEC, which has repeatedly surprised both its critics and supporters with the skill with which it has pursued its own vested interests. In very general terms there does not appear to be any conclusive evidence that would indicate a break in the world price of crude oil. At the most what seems likely is a slowing over the next four or five years of the rate of increase in oil prices which are a barometer of all other major sources of energy. This slowing could bring the rate of increase down to 8 per cent or possibly a little lower.

However, by around 1982, when the effect of the Western controlled sources of oil diminishes and OPEC more or less simultaneously reaches a productivity plateau, oil price increases are likely to start climbing again, possibly to the 10 per cent level. In brief, there is very little concrete evidence to suggest that OPEC will not be able to weather the immediate economic environment comfortably without letting slip its effective control over the world oil price level. In the slightly longer term the indications all point to a strengthening of OPEC's position, though the exact extent of the price increases must be very speculative at this point. None of this is to suggest that the outlook for the oil companies downstream (certainly outside the United States) is anything but fairly bleak; profitability in Western Europe is especially low but such fluctuations do not constitute sufficient evidence of the beginnings of an imminent collapse of OPEC.

As year by year OPEC confounds the soothsayers of collapse, the precise reasons for its survival have been less carefully scrutinised than its long heralded downfall, perhaps because many of the most pressing problems in the world economy are, as we have noted, more amenable to solution if such a price breakdown ensued. There are many reasons that might be deployed for arguing the continuing survival of OPEC for the foreseeable future; the following six are put forward not as a comprehensive answer to those who foresee a collapse but as a reasonable counter hypothesis on a subject upon which no man living can be absolutely certain.

First, the current price weakness in oil products on the market is not sufficiently severe to warrant the expectation that it represents the prelude to the collapse of crude prices; cycles in refined products are normal in the oil industry and occur for very good reasons such as weather variations, overall economic slowdowns and speedups, by inventory stocking of oil before crude oil price rises, etc. Cycles at the market end must always be differentiated from fundamental issues of supply and demand which by their very nature are both much more severe and more sustained.

Second, a great deal has been made of OPEC's spare capacity which is interpreted, if not explicitly then implicitly, as something that OPEC cannot afford. The question is whether it is in reality burdensome. In 1977 OPEC had an estimated production capacity of roughly 40 million barrels a day whereas its actual production during 1977 was not much more than 30 million barrels a day. Thus during 1977 OPEC was operating at about three quarters capacity. This compares favourably with the situation in 1974 and early 1975, when production was not much more than 25 million barrels a day compared with a capacity of 38 million barrels a day. If OPEC production for the immediate future remains in the range of 29–31 million barrels a day, as it can reasonably be expected to do probably up to about 1982, then the adjustments required by most OPEC countries, or at least OPEC as a whole, are far from being too burdensome for it to handle.

Third, the spare capacity which undeniably exists and undeniably creates very real problems for some OPEC members is, for the most part, concentrated in the hands of the 'strong' members. The 'strong' members fall into two categories: those such as Iran, Libya and Nigeria who have fairly consistently pressed for raising prices, and the second and even more influential category represented by Saudi Arabia, Kuwait, Abu Dhabi and Qatar, who have a sufficiency of wealth already so that they can easily do without any additional funds. Apart from the possibility of a price break arising from some major cataclysm such as another Middle East war or a deepening world economic depression along the lines of the 1930s, the probability of a price break originating from the 'strong' group seems remote, given their respective interests and track record to date.

Fourthly, there is no dispute about the fact that Saudi Arabia as the largest single producer and the country least in need of production profits is the lynchpin of OPEC. Moreover, it is a distinctly powerful lynchpin, so powerful indeed that should any one or more OPEC member embark on a major discounting programme, the Saudis would be both willing and capable of adjusting to that fact. Assuming such discounting arrangements were made clandestinely, the likelihood is that Saudi political and economic muscle would hardly need to be flexed to bring any unauthorised discounting member into line. In 1977 the production level of Saudi oil was running in the vicinity of 10 million barrels a day; by 1978 the official target was 8.5 mbd. It is believed that Saudi Arabia could cut back production to 5 million barrels a day and still operate profitably and meet its commitments in full; such, at least, was its bargaining strength. Less disputable and possibly quite as significant is the Saudi policy, adopted by OPEC, toward the apparently endemic problem of world inflation, which in its essentials has been to maintain the price of oil in real terms by making gradual upward adjustments for assumed changes in world inflation rates. While the precise rate of adjustment in future years cannot be calculated, an annual range of 5–8 per cent might be a reasonable figure, given the general expectation that world inflation rates look like moderating in the late-1970s.

Fifthly, OPEC, and Saudi Arabia in particular, is not only capable of absorbing modest discounting of crude oil prices by a dissenting minority; it has in fact always done so, albeit discreetly. Iraq, for example, has for years been moving large amounts of relatively over-priced Kirkuk-Mosul crude to the Mediterranean buyers via the medium of a complex system of pricing formulas which added up to a well disguised discount programme. Indonesia has been in a similar position for some time. Both countries' arrangements have been known and accepted by the other members of OPEC, underlining one of the paramount features of OPEC which seems to have been studiously ignored by some commentators, namely that it is as much a political as an economic entity. The very same signs which have been interpreted repeatedly as omens of imminent collapse have in fact been certain signs of its technical flexibility and its political give and take for which most of the credit must be given to the leadership of Saudi Arabia, which has reinforced its powerful economic position with political wisdom and measured restraint. Without Saudi leadership it is highly doubtful whether OPEC would have survived the internal strains and external stresses of the years since 1973.

Sixthly, probably the greatest single guarantee of OPEC's survival of the period of the immediate oil glut is paradoxically the policies of the main oil consuming countries which are without exception premised on the prospect of higher energy costs. Thus each of the United States, the Soviet Union, Britain, Canada, Australia, Norway and South Africa have embarked upon vast capital investment energy programmes. Moreover, should there in fact occur a significant price break, it would probably not at this stage alter the overall energy stance of the major consuming countries such as the United States, whose basic requirements would be unaffected. While such a price break would manifestly reduce the monetary incentives to produce more energy domestically, simultaneously stimulating the growth in oil imports, it must be doubted whether the US government would return to 'cheap' energy imports given the predicted shortages by the early-1980s. Even with a price break it would remain government policy not only to maintain relatively high energy prices but to seek to introduce modest price increases by gradual means.

Again in the case of the most influential consuming country, the United States, the irony that is becoming apparent is that the Carter Administration runs the danger of reinforcing OPEC by downgrading serious efforts to achieve near-term oil independence in the event of a cut-off. It runs this risk perversely by its initiation of an expanded crude oil stockpiling programme at the very moment when world demand for crude might in normal circumstances be noticeably slackening. This programme of increased stockpiling, far from being confined to the United States, embraces eight major consuming nations in a storage undertaking which, while designed to forestall a repeat performance of the events of 1973–4, can also only serve to undergird for some years to come the current OPEC pricing pattern.

It is against this general background that the particular national energy policies have to be drawn up, neither paying too much attention to the short-term pressures nor over-estimating the inevitability of certain long-term trends but balancing out the various forces at work and reconciling technical feasibility with political imperatives. That the means to justice is closely allied with a framework conducive to competitive efficiency is one of the themes that we will seek to explore in the pages that follow. That conversely, the structures which place a premium on political coercion will not serve the purposes of either justice or efficiency in the field of global energy will be an aspect we hope to prove difficult to refute once the evidence of this book has been surveyed. The prospects for global energy therefore are not simply to accept the trends which have been summarily traced in this very brief introduction but depend to a high degree upon the imagination, wisdom and will applied to the formulation of coherent national energy policies among the major Western industrial countries. On the success or failure of these same energy policies will rest much of the future happiness and prosperity of the Western industrial countries, not to mention the less developed countries, during the next quarter of a century.

PART ONE

THE PATTERN OF PAST POLICIES

1 The Role of Energy Policy

Before the advent of twentieth century industrialism with its previously unforeseen levels of energy consumption, the necessity for a centrally conceived and co-ordinated national energy policy was not generally apparent. By the beginning of this century Communist theoreticians in Russia had conceived of energy policy as one of the principal instruments in creating a socialist society. Its central importance in the structure of command economies has not faltered since that period as successive socialist societies have asserted their control over this vital artery in the industrial state. Unsurprisingly, it has become closely associated with an emphasis on central planning. Yet the evidence now available, corroborated by the evidence of the next chapter, suggests that all contemporary industrial societies have adopted national energy policies. The vital question is have they adopted certain emphases in energy policy with or without forethought of their short, medium and long-term consequences?

1.1 GENERAL ASSUMPTIONS

The experience of all industrial societies throughout the world, but during the last thirty years in particular when modern technology has greatly enlarged man's appetite for energy,

has forcefully underlined that the process of formulating
the broad principles and general direction of national energy
policy is very close indeed to formulating the institutional
framework of the human society we are seeking to create.
If nothing else, energy policy reveals the scope of our expectations. Since 1973 and the abrupt increase in the price
of oil, hitherto the cheapest and most readily available source
of energy in the late postwar era, the awareness that energy
comprised one of the most vital components in sustaining
the pattern of increased material prosperity has become general.
The same dramatic events arising from the application of
the OPEC cartel have manifestly prompted most countries
to reassess their pattern of energy consumption, in effect
to begin rethinking their national energy policies. This rethinking does not presuppose interventionist government policies;
it does presuppose the necessity of taking a great many more
factors into consideration when making decisions in the energy
sector, regardless of whether it lies within the public or private
sectors of the economy.

For what is now generally beyond dispute is that the technology and economics of energy exploitation are going through
a period of very rapid change. These changes, whatever the
benefits or disbenefits they may ultimately confer, create immediate uncertainties and opportunities to the energy planner
of vast proportions which are still being fathomed. The need
for new and more flexible methods of energy planning incorporating regular and repeated critical reappraisals of both the
policies and the principles that undergird them is already
clear. Clearest of all is the fact that energy requirements
will continue to grow and that the means of their deployment
will require ever greater sophistication both in terms of the
technology of conversion and conservation and in the economic
and political structures they are likely to spawn.

1.2   LIMITATIONS

Almost as important as an awareness of the vital character
of the management of our energy resources is an awareness
of the limitations of energy policy. That these limitations
exist and that they will always exist, however sophisticated

energy planning might conceivably become, constitutes a crucial perception. The more diverse the economy, the freer the political and social institutions, the greater the number of variables and the less feasible precise forecasting becomes. It is in the very attempt to predetermine too precisely and too comprehensively the course of long-range energy demand that there lies the greatest threat to liberty. Thus it may be seen that the limits of energy policy are as important as its central objectives and principles. It is nevertheless difficult to lay down limits in the abstract. Recent British experience offers useful examples of the effective limits of energy policy.

The best published attempts at an analytical approach to energy policy emerged in the mid-sixties when the Dungeness B appraisal, the 1967 White Paper and the Report of the Select Committee on Nuclear Power followed each other in quite close succession. The overriding objective of the energy policy which emerged can be summed up in the phrase 'to minimise the costs of energy supply'. To this extent it differed very little from any piece of market research that any commercial organisation might have embarked. The broad strategy advocated, and subsequently adopted by the British Government, was to first estimate the future level of demand and then subdivide this between the direct use of primary fuels and electricity making various assumptions on costs and market preferences. The policy has in retrospect proved to be an inadequate basis on which to formulate energy policy.

The underlying weakness of the analysis was that it was bound to produce an *optimum* policy based as it was on the *best* forecasts of energy supply and demand. In a word this meant that there was too little scope for flexibility, too little room to redirect resources when the optimum forecasts proved unjustified in any particular energy source. The two most notable errors in forecasting were in the costs and availability of nuclear energy and in the price of imported oil which had been assumed (by government apparently, if not the oil companies) to remain relatively stable. While there was always a degree of uncertainty about the nuclear power programme, an uncertainty that is still with us, principally on the grounds that the programme was too ambitious, the central lesson has been the folly of adopting a single overriding aim, that of minimum energy cost. Moreover, the costs of poor forecasting had adverse effects of a wide ranging character. Thus the curtailed investment in coal (representing a rapid decline in production) meant that the shortfall in nuclear power available was met by increased dependence on oil. This manifestly contributed to Britain's chronic balance of payments problem. The price of natural gas may also have been fixed at too low a level if the long term value (i.e. replacement cost) of this finite commodity were taken into account, a point that gets closer and more detailed scrutiny in chapters 5 and 6.

1.3 AIMS

Having suggested the general limits implicit in energy policy, however sophisticated, and of Britain's mid-sixties energy policy in particular, the onus is to suggest an improved basis on which energy policy can be fruitfully formulated. The twin pillars of a reformed energy policy are unquestionably first in the employment of analytical methods which allow for much greater uncertainty in their central assumptions concerning supply and demand in each sector; secondly, that embrace a much broader set of policy objectives than have hitherto been deployed by British energy planners.

The areas where uncertainty in forecasting is likely to persist might appear to some to be self-evident but they require enumerating if only to underline the absolute necessity for flexibility to be built into the central elements of any national energy programme. The areas of uncertainty will include three principal categories:

(1) the rate of economic growth and its influence on energy policy;
(2) the rate of development of both new technologies such as nuclear power or new means of coal conversion;
(3) the future availability and costs of materials, manpower and capital.

These three categories, covering as they do the full scope of the economic environment in which energy production and consumption takes place, will tend to reinforce the sceptics who believe that energy policy really has no place. Without suggesting that a healthy degree of scepticism about forecasting is not entirely necessary and valuable, it is when the objectives of energy policy are seriously debated that the necessity for an overall rationale for energy decisions becomes apparent. The first thing to be acknowledged when attempting to formulate a comprehensive yet flexible series of objectives for energy policy is that by discarding the aim of optimising a single objective, it becomes necessary to reconcile the *conflicting* requirements of several objectives. This becomes a far more demanding exercise than hitherto, but its very complexity represents a far more realistic reflection of the actual forces at work in society affecting energy supply and demand. A conventional list of minimum objectives or aims for energy policy would need to include the following:

(1) to minimise the cost of *long-term* energy supplies; (ie over a sustained period)
(2) to minimise the waste of primary energy;
(3) to minimise the negative effects on the environment;
(4) to minimise the rapid consumption of *premium* fossil fuels;
(5) to minimise the reliance on unproven new technologies;
(6) to minimise dependence on 'insecure' energy supplies;
(7) to minimise capital expenditure both in total and in concentration in any individual sector;
(8) to minimise the dislocation of manpower unless the other factors were conclusive in the long term.

It would be possible to extend such a list very readily but for our purposes it raises the possible aims that need to be incorporated into any ultimate evaluation of alternative energy strategies.

1.4 CRITERIA: FIVE PRINCIPLES

The preceding list of aims give some idea of the range of objectives that need to be reconciled in any durable and effective energy policy. Moreover, in order to assess the possibilities of reconciliation within the framework of specific alternative energy policies there are arguments for formulating comprehensive models of supply and demand (possibly even stronger arguments for very simple models). Such models' chief value is likely to be in the provision of a panoramic rather than a precise picture of how demand might conceivably be met from both primary and secondary fuels. They should also provide an evaluation of the consequences of a particular set of priorities on the demand for investment, the level of energy imports and the efficiency of energy use overall. The complexity of the interaction of various fuels on their respective supply and demand through relative prices, consumer preferences and surpluses and shortages, would become increasingly apparent. All these features would attend any attempt to grapple with any comprehensive national energy plan, in the process deploying specialists as wide ranging as economists, model builders and fuel technologists, together with the active participation of both the policy makers and appropriate administrators. Yet when all has been said such models have a strictly limited usefulness.

It should be noted that comprehensive energy models have not so far proved themselves and the use of a multiple scenario approach — cyclical scenarios for the next five years, trend scenarios for the next twenty-five years — probably offers a more flexible and realistic if more modest solution. Meanwhile the essential criteria which must characterise any energy policy might well comprise the following five principles:

(1) the price of fuel over a relatively sustained period;
(2) the convenience of a particular fuel, including their transport and storage etc;
(3) the security of supply: this may involve either complete self-sufficiency or at least diversity of supply;
(4) the employment of a judicious choice over the conservation of fuels which have a premium usage;
(5) the pursuit of policies that will not only minimise pollution, etc, but erect checks against the spoliation of the future environment.

It is noticeable that of the five principles listed probably only the first two featured as major considerations in energy policy of a decade ago. The importance of the latter three principles is likely to increase rather than diminish. Moreover, the incorporation of the latter principles will effectively internationalise most national energy policies since even the means to self-sufficiency requires a sharing of technological know-how between states; likewise environmental protection techniques, while conservation of particular fuels implies more efficient conversion processes for some of the others. In a word, the adoption of such principles as a minimum starting point upon which to build an energy policy is a tacit admission of the interrelated and interchangeable characteristics of energy in an industrial society.

1.5 ECONOMICS AND TECHNOLOGY

The major role or contribution of a successful energy policy is that it provides a broad framework in which the competing attractions of various energy sources (together with their broader societal consequences) may be roughly estimated. It must be emphasised that it provides no more than a framework and any attempt to maintain an inflexible dependence on its guidelines can be disastrous when forecasts prove manifestly inaccurate. The requirements of any energy policy in a nation relatively well endowed with domestic energy resources is likely to be more contentious than those of nations who possess scarce indigenous energy resources — if only because the energy sources are in competition for the patronage of consumers, who normally want the cheapest fuel available, and of government if they are state enterprises, for obtaining the greater share of investment.

Assuming that energy policy has a useful contribution to make in the monitoring of likely consequences as well as a very loose regulation of an industrial economy (as distinct from a means of dictating its precise direction), there remains, even in such a minimalist approach which this author favours, a fundamental tension that tends to arise within most energy policies between the claims of economics, or the market, and those of technology. This conflict is much more theoretical than practical since both factors must of necessity recognise the importance of the other element. Historically, however, economists will by instinct and necessity, because theirs is a quantifiable approach, conduct the debate over priorities on the basis of known reserves and capacities etc; technologists and scientists, on the other hand, have a counter-balancing tendency to argue for the vast future potentials, if technology is applied in concentrated form in the appropriate sectors, ie those nominated by the technologists. There can be no automatic acceptance of either approach but what is clear is that the element which characterises successful national energy policies is when their respective approaches have been reconciled and incorporated into the balance of gain and loss.

The clearest example of this element in the debate over British energy policy would be the arguments deployed by the coal technologists that the future for new coal conversion processes is so promising (and the reserves of coal so extensive) that investment in conversion processes should be greatly accelerated. The adoption of this policy, involving the creation of coalplexes, and retaining the current state control over coal, would have the effect of encouraging the integration of energy industries, which would in turn reduce the degree of competition that currently still exists between the various nationalised energy industries (besides those in the private sector).

It is a major development, ie the various coal conversion processes, that highlights the built-in conflict between the economist and technologist, a debate that is an even hotter issue in the field of nuclear energy. For those who sometimes suspect the scientists of over-arguing their case, Britain's experience, indeed Western Europe's generally, over nuclear energy during the last decade provides a salutary warning against constructing too substantial an economic edifice on ground which was scientifically unproven. It is a warning that might be as appropriate in the fields of alternative energy sources generally, as well as the nuclear sector whose much vaunted claims of a decade or so ago have induced scepticism about its future contribution.

2 International and Comparative Energy Policies

It is the besetting sin of national energy planners to underestimate the degree of interdependence of the world energy economy which persists despite the strenuous efforts and ambitions of government programmes designed to create regional, if not national, energy self-sufficiency among the major industrial powers. The only antidote to this phenomenum is to stand back and survey the world energy economy as a whole and then go on to attempt to summarise the energy policies of the major powers who constitute by far the greater part of world energy production and consumption despite the crucial part played by the OPEC states in the world oil market.

2.1 GLOBAL ENERGY

The first and most striking feature about the global energy picture is the extraordinary rapid postwar growth in overall consumption, a growth pattern which looks likely to be maintained for the indefinite future. Whereas in 1950 world energy demand amounted to 2600 million metric tons of coal equivalent (hereafter called *mtce*) comprised of 1600 million tons of coal, 700 million *mtce* of oil, 260 million *mtce* of natural gas and 40 million *mtce* of hydro-power, by 1972 total world energy consumption had soared to 7600 million *mtce*, of which nearly half, that is 3350 million *mtce*, was accounted for

TABLE 2. *World Energy Demand to 2020*
Energy Projections 1972–2020
Energy in ExaJoules (10^{18} J)

	Energy Constrained High Growth H5			Oil Constrained Low Growth L4	
	1972	2000	2020	2000	2020
Secondary energy demand					
Fossil fuel	156	306	440	259	338
Electricity	17	58	120	54	109
Wood, Solar, etc.	31	59	82	55	75
Total Secondary energy	204	423	642	367	523
Primary energy demand					
Coal	66	143	361	125	278
Oil	114	195	194	159	135
Natural Gas	46	71	88	66	85
Nuclear	2	90	231	83	212
Hydro	14	34	56	34	56
Wood and Solar	26	52	75	48	69
Total Primary demand	268	584	1005	515	833
Total production	274	636	953	606	924
Net imports	–(6)	–(52)	52	–(91)	–(91)

Primary energy demand	Unconstrained High H3			Unconstrained Low L3	
Unconstrained H3 and L3	268	730	1526	531	899

Primary energy demand	Normalised with 1972 = 100				
Unconstrained H3 and L3	1.0	2.7	5.7	2.0	3.4
Constrained H5 and L4	1.0	2.2	3.8	1.9	3.1
Total GNP	1.0	3.2	6.8	2.4	4.2

Primary energy demand	Average annual percentage growth			
	1972–2000	2000–20	1972–2000	2000–20
Unconstrained H3 and L3	3.6	3.8	2.5	2.7
Constrained H5 and L4	2.8	2.8	2.4	2.4
Total GNP	4.3	3.8	3.1	2.9

by oil. The expectation in late-1973 was that world energy
consumption would reach around 16000 million *mtce* by 1985
at the most conservative estimate. These summary figures
underline the extent to which vast international energy consumption is a vital aspect of postwar industrialism. The dramatic increase in world consumption can be broadly accounted
for by both the spread of industrialism and an increased

rate of per capita consumption as part of the growth of a technologically sophisticated form of industrialism. In addition there has been a manifest growth in total world population, though this has been largely in low energy consumption areas so its contribution to increased world energy consumption should not be overestimated.

The second, no less striking, feature of world energy consumption was that, by 1970, the industrialised world obtained 94 per cent of its total energy requirements from fossil fuels. This pattern, which followed the lead of the United States (which had long relied upon coal, oil and gas), was in stark contrast to the non-industrialised countries, representing 70 per cent of the world population, who remained heavily reliant on traditional 'local' energy sources such as work animal feed, fuel wood, windpower and direct waterpower. The published findings of the Ninth World Energy Conference, held in Detroit in 1974, cast an interesting light on the precise breakdown of the world's estimated recoverable energy reserves. The conference estimated that the total resources to solid fuels was 11 million million tons giving a resource to demand ratio, i.e. a life expectancy, for solid fuels of 3760 years. The solid fuel resources classified as reserves amount to 1.5 million million tons, which equals a reserve to demand ratio or life expectancy of 500 years. Within this overall estimate of reserves there is a further sub-division of reserves which are currently considered recoverable, given as 0.6 million million tons with a reserve to demand ratio of 200 years. The view of the World Energy Conference in 1974 was that reserves in the latter category were most likely to remain stable, for the next few decades at least, as new economic reserves are generated.

A third remarkable feature of world energy the Conference revealed was the breakdown of recoverable reserves as between solid fuels and other sources of energy. This included the surprising result to some that (discounting the potential of breeder reactors, whose technology was still unpredictable in performance) around half of world recoverable reserves are in the form of solid fuel. This only underlines the fact that coal represents by far the largest share of fossil energy in the world. The same WEC survey suggests that the reserve to demand ratio is broadly uniform over the major world regions, though with certain notable exceptions; they include Japan with three years, France with five years and Israel with four years. Allowing for the imprecision of such calculations, it confirms the general vulnerability and urgency of energy policy to each of these three highly industrialised nations. It is no accident that, as we shall see later in this chapter, Japan has embarked upon an ambitious programme of resource diplomacy to enhance her energy supplies while both France and Israel have sought the fastest possible development of their nuclear energy sector that is consonant with their financial resources. Britain's reserve to demand ratio is calculated at a conservative twenty-one years, a comparatively favourable prospect by international standards.

A fourth and final general feature of world energy is its heterogenity, meaning that among the various categories of energy each has its distinctive usefulness in special spheres, a factor which needs incorporating into future energy demands. The newer — what are bracketed in this book as alternative energy sources, which include, solar, geothermal, hydro and tidal energy — represent the hetorogenity of energy sources in the long term. The impact of the alternative energy sources before the end of this century is not likely to be substantial for a variety of discernible reasons. Generally speaking they have three main shortcomings: first, they provide only low grade heat; second, they demand as yet an inordinately high level of capital investment; and third, they are frequently located some distance from the ultimate centres of consumption, greatly increasing the storage and transmission costs.

Despite the heterogenity of energy supplies theoretically capable of development and because of the close relationship between energy, economic stability and prosperity — and even social welfare within the current industrial structures — overall energy costs are likely to rise steadily with energy consumption remaining, on the long haul at least, relatively insensitive to price in the sense it does not historically seem to act as an effective brake on consumption. This does not, of course, mean that price will not help determine the particular energy source consumed, only that it does not apparently dictate overall energy consumption levels by itself.

Nuclear power, in particular, is a good example of a fuel likely to become increasingly rather than decreasingly expensive to produce, most notably because the environmental safeguards will become much more numerous and their legal back-up more enforceable as public opinion realise the inherent risks. The environmental factor will also be increasingly taken into account in the prospecting for coal and oil, significantly adding to their production costs without taking any account of the likely impact on costs of temporary or even sustained shortages of particular fuels. One possible restraint on the trend toward energy prices rising would be the successful commercial intro- duction of breeder reactors, stabilising fuel costs for burner reactors. However, they are enormously capital intensive in the short to medium term and will only conceivably come into the reckoning in Britain, for instance, by the very end of the twentieth century.

2.2 COMPARATIVE ENERGY POLICIES

When it is remembered that the United States and the Soviet Union are the world's main producers and consumers of energy and that together with Japan, the European Economic Community and China, they account for the major share of world energy consumption, with the United States alone consuming very nearly a third of total global energy consumed, it becomes apparent how central to the course of the world economy in general (and the world energy economy in particu- lar) the energy policies of the five major powers have gradually become during the last decade. Both because it is at once the largest consumer of energy by far and at the same time potentially among the richest sources of energy of all sorts, and because it has declared its willingness to share with other nations the results of its research and development in energy, it is reasonable to begin with outlining the recent energy policy of the United States.

2.3 THE UNITED STATES

Against the background of accelerating rates of consumption, it has been clear from the very early-seventies that the United States would exhaust its supplies of domestic petroleum by the mid-1980s unless fundamental changes were made in national energy policy. Since oil represented 45 per cent of US energy needs and the nuclear programme was nowhere near advanced enough to take up the slack, it was apparent long before the 1973 Middle East war broke that drastic changes were required. In 1971, President Richard Nixon, addressing the Congress, urged the expansion of research and development on energy, a trend which undoubtedly began to improve in the next two years with expenditure on research and development increasing by 50 per cent. This was, however, nothing like an adequate response to the situation which was becoming painfully obvious. Namely, that the United States, was becoming daily more dependent on oil imports, more especially from the Middle East, where 75 per cent of the world's proven reserves of oil lies along the shore and seabed of the Persian Gulf. The likelihood of Arab pressure against Israel through the cessation, restriction or price inflation of Arab oil supplies, together with an implicit potential threat to the entire industrialised world through the international money markets, was early foreseen as a chastening possibility. By January 1973, in the course of a US Senate enquiry, top Pentagon officials revealed that in 1972 the Defense Department was obliged to purchase nearly 50 per cent of its fuel from abroad, dramatically underlining America's vulnerability. This crucial piece of evidence spurred on those in favour of encouraging exploitation of domestic energy resources, notably the North Slope of Alaska, and arguably clinched the future commitment of the federal government to the goal of energy self-sufficiency as soon as it was politically feasible, that is in 1973.

With the sobering thought that the United States energy requirements would by 1985 be double what they were in 1970, President Nixon revealed in April 1973, a five-point action programme which provided the basis for fundamental US energy policy since that date. The Five Points included the following objectives:

(1) to increase domestic production of all forms of energy;
(2) to act to conserve energy more effectively;
(3) to strive to meet energy needs at the lowest cost consistent with national security and the national environment;
(4) to act in concert with other nations to conduct research

in the energy fields and to find means of preventing serious shortages;

(5) to apply vast scientific and technical capacity — both public and private — so that the United States might use its current energy sources more wisely and develop both new sources and new forms of energy.

To ascertain how the broad strategic principles have been translated into practical policies, or not, it is essential to establish the central characteristics of US energy supply as it currently operates. At present over 90 per cent of US energy comes from three sources — natural gas, coal and petroleum; their respective advantages are for natural gas, that it is the cleanest, but that it is also the most scarce; for coal, that it is the most plentiful, but that it creates the worst environmental problems; for oil (formerly both cheap and plentiful), that with domestic production no longer capable of meeting demand it will gradually become restricted and expensive.

Perhaps the most important single set of statistics giving a clue to the balance likely to be struck between the major energy sources was the announcement by then Interior Secretary, Rogers C. B. Morton, when promoting stepped-up strip-mining legislation, that America's ground energy reserves consisted of 4 per cent in oil, 3 per cent in natural gas and 91 per cent in coal. While these figures hide the still unexplored potential of America's offshore oil and gas, apart from the nuclear energy programme and the various alternative energies, coal must obviously play a substantially more important part in US energy policy.

The central umbrella under which US energy policy has been nurtured since 1973, bringing together a number of policies which had been evolving for some time, was labelled by the Nixon Administration, Project Independence. By concentrating the nation's efforts on the exploitation of America's vast coal reserves, by deregulating natural gas (ie by allowing the price to rise providing the incentive for further exploration), by bringing forward the nuclear power programme and the supply of oil from the Alaskan fields, the aim of Project Independence was to make the United States free from dependence on foreign energy sources by 1980. These objectives

have proved more illusory and less easily attainable than
was first thought and the goal of self-sufficiency may not
even be realised on current performance by the mid-1980s,
let alone 1980.

By 1975 President Gerald Ford, in an attempt to cut the
consumption of oil in general and imported oil in particular,
proposed a plan to increase the price of imported oil by
three US dollars per barrel, a proposal that Congress rejected.
Undeterred, President Ford pressed for the decontrol of the
price of 'old' domestic oil (old oil is defined as that produced
from wells existing in 1973 at a rate equal to 1972 production)
and by December 1975, had reached a compromise with
Congress which essentially allowed controls to be retained
for an interim period with the prospect of eventual decontrol
being introduced.

Natural gas so far evaded a similiar compromise arrangement
though its importance cannot be underestimated. Against the
background of a shortfall in natural gas available, and a
declining rate of production, the Ford Administration asked
Congress to end twenty years of federal control on prices
paid at the wellhead for America's cheapest and cleanest
fossil fuel. Opponents of deregulation argued that it would
place the consumer at the mercy of the oil companies who
were the major producers of natural gas. Critics further argued
that deregulation would merely create windfall profits without
guaranteeing that more gas was produced, an argument which
the President quickly countered with a windfall profits tax
proposal.

The progressive deregulation of gas passed only recently by
the Congress after a protracted struggle serves to underline
the central importance of gas. If oil imports are eliminated,
natural gas becomes America's principal current source of
energy. Among the chief sources of US domestic energy are
natural gas 41 per cent, crude oil 30 per cent, coal 22 per
cent and 6 per cent from hydro and nuclear power. Even
with deregulation, the Project Independence report anticipated
that by 1985 gas production would have reached a level only
slightly lower than the level in 1975, far below the requirements
of the mid-1980s.

Certain features stood out in the overall shape of US energy
policy by the mid-1970s. By setting a target of eliminating
US dependence on foreign energy within a ten to fifteen

year time span, Project Independence implicitly accepted increased costs of about one third for consumers. This in effect was never accepted by Congress, whose outright rejection or compromise measures in response to the Nixon and Ford Administrations's proposals have left Project Independence virtually high and dry. The consequences by the end of 1976 were far from promising; with domestic oil output declining, a deteriorating balance of payments deriving from increased oil imports (then 40 per cent of total), a rising inflation rate partially engendered by the growing deficit on the balance of payments to say nothing of the menacing shape on the horizon of a further substantial oil price rise on the part of the OPEC states.

The apparent prospects for 1977 were no better with Congress seeming to be preoccupied with divestiture, that is the breaking up of the largest of the oil companies. The most probable response on the part of the major companies then looked like retrenchment and diversification into other, not necessarily related activities. Meanwhile the new President, although empowered by the Standby Energy Act to deal with any new OPEC embargo, is still not able fully to resist or contain a well-timed and co-ordinated oil price rise. Finally, as far as US energy policy is concerned, as long as Congress was so preoccupied with divestiture, it looked unlikely to embark on a comprehensive, balanced programme to both conserve and develop alternative energy sources, an ideal to which it has publicly committed itself within the framework of the International Energy Agency.

2.4 THE SOVIET UNION

The energy policy of the Soviet Union, like those of its overall economic policies and unlike those of the Western industrial powers, has always been aimed at ideally national, or at the very least regional, self-sufficiency. With the increasing technological sophistication of the Soviet economy, accompanied as everywhere else by accelerated rates of energy consumption, this has been a goal progressively more difficult to accomplish. Moreover, in its unswerving priority to the maintenance and expansion of its military establishment, the Soviet Union has of necessity failed to create the kind of economy which would generate the broadly based technology characteristic of the United States, Japan and the European Economic Community. By its lack of such technology, the Soviet Union has to some extent itself undermined its economic self-sufficiency, and especially energy self-sufficiency, which its perennial war economy demands as a logical economic back-up.

The principal response of the Soviet leadership to this threat to the traditional policy of energy self-sufficiency has been the so-called Siberian policy. Although the beginnings of the Siberian policy can be traced as early as the late 1950s when the Sino-Soviet conflict first surfaced, when it represented an historic embarkation on the economic and industrial development of this hitherto neglected harsh hinterland of the Soviet Union, it was not long before the discovery of vast mineral and especially energy resources transformed its significance from one of strategic regional importance to a central plank in Soviet economic and defence policy. From a military standpoint the Siberian policy represented a concerted effort to transfer a significant proportion of Soviet productive capacity to the East, thereby bringing them closer to the major future sources of raw materials and fuel while at the same time dispersing industrial targets against external attack from the air. Economically and industrially there can be no doubt that Siberia represents the greatest treasure house of Soviet energy and raw materials currently known to exist; not only does it represent the key to Soviet supply and demand outlook for energy, but it also has worldwide ramifications, representing as it does a major share of global energy resources and world trade in energy in addition.

By 1974 the Soviet Union became the world's largest producer of oil when its production of crude reached 3400 million barrels, around 200 million more barrels than the United States. However, oil production targets have recently been met only by the excess production of one large field. Soviet authorities claimed oil resources already prospected were sufficient to last for several decades and in addition asserted they possessed one third of all global deposits of oil and gas. But the really important question that remains unanswered is not the extent of Soviet reserves of oil, gas and coal, which are almost certainly gigantic by world standards, but the capacity of the Soviet Union, first, to meet her national energy requirements from domestic sources for the short-term future, and second, to obtain or develop her own technology to extract the highly inaccessible Siberian energy resources. The Soviet Government's policy remains one of total energy self-sufficiency, but it may yet have to settle for diversity of supply, at least for the medium-term future.

If one were to treat Comecon, that is the Soviet Union and what are mostly her East European satellites, as a self-sufficient entity, then in very broad terms the Soviet Union would appear to be self-sufficient already in not only energy but in most of the major raw materials demanded by an industrial society. The great exception to this self-sufficiency is in agriculture where the Soviet Union's grain shortages are well known. There is nonetheless a distinct possibility that the Soviet Union may need to import oil on a growing scale in the early-1980s. By comparison with the continued dependency of the major Western industrial countries on OPEC it is a minimal degree of dependence.

2.5 THE EUROPEAN ECONOMIC COMMUNITY

It should be emphasised that the following section represents a summary of the evolution of an embryo EEC Common Energy Policy, a record devoted as much to the problems lying behind the slow progress made toward an effective policy as to any striking achievements in the formulation and operation of such a policy. A factual record of the state of energy supply and demand and the effects of the interaction of the various national energy policies within the Nine follows in a brief third chapter.

The slow evolution of an EEC Common Energy Policy should surprise nobody. The reasons are manifold and might be described as part historical and part structural (ie they are tampering with one of the tap roots of not only national sovereignty but the independence of some of the most powerful elements in the structure of industrial society). Historically, the slow development of a common energy policy was due to three principal factors.

First, responsibility for energy was divided between the various organs of the Communities: the Paris Treaty conferred responsibility for coal on the European Coal and Steel Community; the Rome Treaty assigned oil, natural gas, electricity and hydro-power to the EEC Commission, and left nuclear power development and control to Euratom.

Second, at the outset of the Community there was no mention of a common energy policy for the fairly simple reason there was no apparent need to co-ordinate differing energy sources since coal was still king. In 1950, for instance, coal provided 75 per cent of Community energy requirements, oil a mere 10 per cent. By 1966 coal had dropped to around 40 per cent with oil climbing to about 50 per cent these changes were almost entirely wrought by the overall growth in energy consumption with the additional demand for cheap and immediate energy between 1960 and 1970 being met almost exclusively by oil. Overnight the Community had become the richest oil importer in the world for which the multinationals provided a cheap and plentiful source of energy throughout the 1960s.

The third historical but also very contemporary reason why the common energy policy has made such slow progress is that, in order to come into full operation, it would need to cut across national policies, nationalised industries and fiscal policy; in short, to challenge the national energy interests across the board. This has been the most fundamental reason for moving slowly toward a common policy, the more so because different sources of energy are not of equal importance in each country. Thus Italy, the Netherlands and Luxembourg, with only small or non-existent coal outputs, for long favoured a cheap fuel policy, effectively supporting the prevailing trend toward imported oil. By contrast, each of West Germany, France and Belgium operated major coalfields, with West Germany in particular providing 75 per cent of the Six's coal requirements. Their arguments for self-sufficiency, effectively an argument for increased reliance on coal, were overruled by the availability of cheap imported oil. It is interesting to speculate that if France rather than the newly emergent West Germany, still seeking to establish its political acceptability, had produced the majority of the Community's coal, whether coal might have been aided to hold its own against the challenge of oil in the 1950s and early-1960s.

The dramatic consequences of the OPEC cartel enforced on the Western industrial nations in late-1973 was bound to act as a goad on the formulators of the EEC Common Energy Policy. As early as 1971 the energy companies warned the EEC Commission of an impending oil crisis. It was nevertheless not until December 1974 that the EEC Council of Ministers was able to agree on a package of conservation and diversification measures to reduce energy consumption growth by 15 per cent, and dependence on imported energy from 63 per cent in 1973 to between 40 and 50 per cent in 1985. Previous to this the Commission had reasserted that any strategy for Community energy policy would need to observe three basic criteria: first, to maintain price levels to the consumer as low as possible; second, to ensure reasonable profitability to guarantee sustained investment; and third, to create the framework for greater conservation by consumers. Although, like the Nixon Administration's Project Independence, the overriding long-term goal of the Commission was to make the Community self-sufficient in energy, this was not either a short or even medium-term option. Instead, the Commission aimed at two major objectives; first, the guaranteeing of security of supply by means of diversification of domestic and foreign energy sources; and second, the creation of a unified market for energy. Of the two, the first was the more immediately realisable, the second presented a very distant prospect.

To summarise the present state of the attempt to create an EEC Common Energy Policy, it can be said that there exists a two-tier strategy: one for ten years, and the other for twenty-five. Both strategies propose as their major objective the reduction in the Community's dependence on imported oil and the development of nuclear power, coal and gas. The Community's drive toward greater self-sufficiency is, needless to say, heavily dependent on the successful exploitation of the vast new resources of the North Sea, which is why British energy policy is so important not only to Britain but to all the other member states of the European Economic Community. By seeking to describe the attempts of the Community to create a common policy, and also the actual trends in energy supply and demand within the Community's members, the author is seeking to provide the necessary background to understanding the requirements of a British energy policy which harmonises Britain's national self-interest with the national interests of her Community partners.

By demonstrating the basic lack of success of the EEC

Commission and Council in their efforts to forge a common policy, this treatise hopes to argue the case for attaining many of the same objectives of the common policy by a very different route. This route is fundamentally taking the energy policy of the best endowed partner in the Community, namely Britain, creating a rational framework for future development and seeking to recommend that policy as the basis on which the Nine's energy strategy might be formulated.

2.6 JAPAN

Not unlike the European Economic Community, only more so, Japan has become increasingly dependent on imported sources of energy. This import dependence was an integral aspect of the astonishing economic growth of Japan during the 1960s in particular, when imported oil was cheap and plentiful. More than for any other major economic power, the figures make startling reading. More surprisingly still, the general expectation is that, while concerted attempts at domestic diversification of energy sources will be complemented by diversification of foreign suppliers of both oil and gas, due to the escalator of increased overall energy consumption, Japan will remain predominantly dependent on imported energy for the indefinite future.

The logical starting point for a summary review of Japanese energy policy is the striking sparsity not only of domestic energy resources but of raw materials generally. Japan lacks worthwhile quantities of both oil or coal and among her indigenous sources of energy only those of hydro-electricity and thermal power count for anything at all. Yet Japan maintained an annual rate of growth right up until 1972 of around 12 per cent which she had sustained from about 1960. Unsurprisingly, during the same 12 years her rate of energy consumption, largely imported, rocketed. Thus between 1962 and 1972, total energy consumption grew threefold, with oil and gas increasing fivefold each and coal and electricity by a factor of 1.4.

During the same period the Japanese economy was steadily switching from coal to oil, becoming more vulnerable year by year as it became simultaneously more import dependent.

In 1962 coal accounted for 36 per cent of energy consumption, oil 46 per cent; by 1972 coal had fallen to 17 per cent while oil had soared to an astronomical 75 per cent. The overall comparison of imported energy was 52 per cent in 1962 compared with 86 per cent in 1972. The fact that the great bulk of this imported energy derived from a single region, that is the Middle East, was to make Japan the most vulnerable by far of all the major economic powers when the OPEC countries finally fully exercised their hitherto restrained bargaining power.

The severity of the OPEC challenge to the Japanese government and people was greater than any thrown up in the postwar era, a challenge heightened by its comprehensiveness and its suddenness. Shaken but not intimidated by this challenge, the Japanese government based its counter-attacking strategy on four main priorities: firstly, to economise on oil consumption immediately; secondly, to build new enlarged reserves of oil; thirdly, to seek out fresh sources of crude oil; and fourthly, to reduce and eventually eliminate Japan's current account deficit through a major export programme. With admirable dispatch the government had by November 1973 imposed a 10 per cent cut in consumption of both oil and electric power, passed two major pieces of legislation (one for balancing oil supplies equitably among consumers and the other for stabilising people's livelihoods) and declared a state of national emergency. This was only the beginning and it was not until August of 1974 that the effects of the cumulative governmental measures began to have a generally beneficial impact.

The next step was the inauguration of an energy recovery programme spearheaded by the Ministry of International Trade and Industry (MITI). MITI's programme strategy can be summarised under six main headings: (1) the active development of oil in Japan's offshore continental shelf; (2) the direct acquisition of crude oil from oil producing states on a government to government basis; (3) the increased stockpiling of oil until, by 1979, the nation should possess a 90-day supply in reserve; (4) the readjustment, mainly slimming down, of Japan's energy industries; (5) the development of new energy technologies; and (6) the promotion for the long term of both the nuclear and solar energy industries.

Taken together, MITI plans the reconstruction, or more

precisely, the restructuring of the energy economy so that the growth in oil consumption can be appreciably slowed. The means employed necessarily increased the degree of government intervention. A prime example of this is the government's stockpiling policy which offers loans on very generous terms to the private oil companies to enable them to buy or build new storage facilities and to buy crude for stockpiling. The other notable example of increased government involvement is the so-called Sunshine Project involving government sponsorship in the development of new technologies for using new sources of energy such as solar, geothermal, coal gasification and hydrogen energy. Meanwhile, throughout 1975, the newly created Japanese Advisory Committee on Energy was heavily engaged in securing government-to-government deals in oil and attempting to diversify away from the Middle East, which nevertheless remained Japan's principal regional supplier.

2.7 CHINA

In examining the energy policies of the United States, the European Economic Community and Japan, several common features emerged, among them the continued growth of long-term consumption and the attempt by all three to create if not self-sufficiency then a much greater diversity of supply. China differs markedly from all three in that its rate of consumption is appreciably lower than the previous three on a per capita basis and shows very little immediate signs of accelerated consumption, though the possibility remains that it might do so should the post-Mao leaders of China decide to press forward with rapid industrialisation deploying Western technology rather than follow their now traditional policy of self-reliance. The development of China's energy policy is profoundly revealing about the course of Chinese society in the postwar period.

In the early stages of the Communist government, that is during the 1950s, China embarked upon a programme of industrialisation similiar to that of its political and economic sponsor, the Soviet Union. This concentration on heavy industry involved a significant increase in energy consumption which in turn produced a heavy reliance on imported Soviet energy and raw materials. By 1959, as part of the deteriorating relationship between China and the Soviet Union, the Chinese Communist Party Central Committee made the momentous decision to drastically modify the Soviet-style virtually exclusive concentration on heavy industry and concentrate instead on improving agriculture.

As part of this new emphasis on agriculture, the Central Committee of the Chinese Communist Party decided upon rural electrification over a target ten year period. This effectively meant that hydro-thermal power was to be given top priority at a stage when energy resources were far from fully explored, let alone exploited and developed. It was already apparent, nevertheless, that China's four main sources of energy in order of importance were coal, water power, petroleum and natural gas. In the dramatic recent discoveries of oil, with its potential as a major export earner, and the intriguing possibility that the new Chinese leadership might eventually adopt policies of Western style industrialisation, it would be easy to overlook the true nature of China's principal energy sources.

Though their development has been markedly more gradual than oil, coal and water power remain China's chief energy sources, especially coal. China's reserves of coal, estimated at one trillion tons make her the third best endowed coal country in the world, following in importance the Soviet Union with 4.2 trillion tons and the United States with 1.1 trillion tons. In China's case her reserves have scarcely been scratched. In second place in her energy larder, China's water power potential was assessed in 1958 at 5800 million kilowatts, utilising one of the two (the other is the Soviet Union's) most extensive national river systems in the world. A brake on the full exploitation of China's water potential, however, is the fact that many of these rivers lie in thinly populated areas of Yunnan and Tibet. Third among her principal energy sources was oil. By 1975 China had become the fourteenth largest oil producer, very largely thanks to a single major field at Taching whose production is still climbing rapidly. Oil, however, still represents only about one fifth of China's energy requirements, which underlines the diversity of China's domestic energy sources and her ability to maintain total energy self-sufficiency, making her the envy of the other major economic powers.

As we shall hope to demonstrate, in its ability to restrain energy demand at home, in its capacity to cultivate a diversity of domestic energy sources and not least in its willingness to export forms of energy for which the external demand is acute, namely oil, and which can be effectively substituted on the domestic market, China provides a model of intelligent and resourceful national energy policy which could well be emulated with benefit by more than one Western industrial country. That it may well demand a degree of restraint, even of austerity, on a scale not previously exercised by many Western industrial societies should not preclude its adoption by countries whose industrial economies are plainly out of balance.

3 Energy Trends in the EEC

In the previous chapter a summary account of the attempts to date to forge a common EEC energy policy was included among the outline energy policies of the major economic powers. In reality there is a glaring gulf between the objectives of a common EEC energy policy and the actual pattern of energy supply and demand within the Community. The aim of the following chapter is to describe the actual pattern of EEC energy trends as they have developed in the past and as they seem most likely to develop in the foreseeable future. Without the provision of such a backcloth in the shape of the political objectives that the central institutions of the Community wish to impose, and the historical record of the energy patterns traced by the Nine member nations, the discussion of the energy alternatives cannot be seen in context.

3.1 TRENDS OF THE SIX

As in all advanced industrial societies, energy has played a central part in determining the character of not only the industrial infrastructure but also the social structures of the Six as well. Prior to enlargement of the Community, the industries of the Six spent around one quarter of their annual budgets on investment in energy; in the iron and steel industry

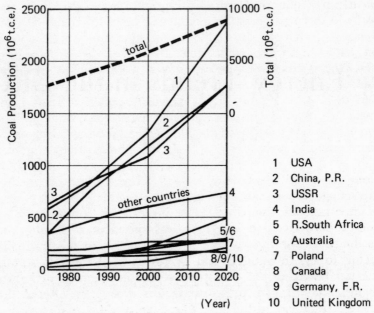

Survey of the future trend in production figures for the main coal-producing countries

the proportion of investment was about 25 per cent; in chemicals, cement, ceramics, glass and non-ferrous metals, around 15 per cent; in the food industry, about 10 per cent. Moreover, the companies engaged in the energy sector are themselves among the largest enterprises operating within the Community — of which more about later in this chapter — where they provide jobs for roughly one million workers and indirectly are responsible for helping to create employment for many tens of millions. It is immediately apparent to even the casual observer that the availability of energy is a prime factor in the selection of industrial sites and thus ultimately in the development or lack of development of particular regions. Even before 1973, energy, predominantly for industrial use, accounted for roughly one fifth of all imports from outside the Community. It is the long term development of Community energy imports that should particularly concern us.

As the Community began to consume increasing quantities of oil, its import dependence grew correspondingly. Thus in 1950 the Six imported 13 per cent of their energy requirements; by 1960 the figure had grown to 30 per cent and by 1970 to 63 per cent. In descending order of energy dependence (1970 figures), Luxembourg was almost totally dependent on energy imports, while Belgium and Italy imported 82 per cent, France 71 per cent, West Germany 48 per cent and the Netherlands 42 per cent. Thus, over a twenty year period, all the original EEC countries except the Netherlands (which discovered domestic sources of natural gas) have become increasingly dependent on outside sources of energy. Importing more than 95 per cent of the oil she consumed, the Community had by the turn of this decade become, at 580 million *tce* (tons coal equivalent), the world's largest crude oil market attracting, as we shall catalogue later in this chapter, a host of international companies. The increase in consumption of energy between 1950 and 1970 was spectacular. Taking 100 as the index for 1950 it rose by 1970 to 264 in the transport sector, to 257 in industry and to 358 in the household sector, the last reflecting the growth in household demand or family living standards. In 1973 two factors drastically altered the energy situation within the Community; the first was the admission of Britain with great reserves of oil, natural gas and coal; and the second was the OPEC oil price rises following in the wake of the October 1973 Middle East conflict.

3.2 TRENDS OF THE NINE

With its high degree of energy import dependence which it inherited from the original Six, the currently enlarged Community of Nine is an integral part of the world energy market even though the impact of that market will obviously vary from country to country within the Community. During the postwar period, world energy consumption has sustained a growth rate of around 5 per cent per annum; this has led to a doubling of demand every fifteen years to which the Nine have been no exception.

Thus in 1972, the year prior to enlargement, the Community's energy consumption was 1260 million tons of coal equivalent or around 16 per cent of world energy consumption. In per capita terms this represented 4950 kilogrammes coal equivalent which is about two and a half times greater than the world average. Prior to the 1973 energy crisis, the EEC authorities believed that its energy demand would continue to rise at 4.7 per cent per annum until 1985, giving a consumption total of 2575 million tons of coal equivalent in that year. In fact, due to the economic recession in the Community largely brought on by the oil price rise, there was an overall decline in energy consumption of 4.7 per cent in 1975 as compared with 1974 which had registered a sharp decline from 1973. Within the 1975 total there were wide variations; consumption of oil and coal fell by over 8 per cent, while largely due to increased availability as well as price competitiveness, consumption of natural gas rose by 7.8 per cent. The decline in the consumption of energy in 1975 in the Community overall reflected fairly accurately a fall of 7 per cent in the industrial production of the enlarged Community.

The modest economic recovery in the Community in 1976 — the gross domestic product rose 3 per cent compared with 1975 — was matched by a roughly similar increase in energy consumption in percentage terms. Translated into *mtce*, this meant an increase from 874 million tons of coal equivalent in 1975 to 900 *mtce* in 1976. This was still 4 per cent lower than 1973 (936 *mtce*). The inland consumption of primary energy within the Community for 1976 reveals the greatest increased demand being for natural gas and electricity. In 1976 the percentage consumption of energy, in order of importance, was oil 54 per cent, coal 20 per cent and natural gas 17 per cent, together accounting for 90 per cent of energy consumed within the Community. These figures underline the marginal importance at this point which the alternative energy sources are capable of making in the immediate future when even nuclear energy has some way to go before it is making a contribution on the scale its proponents recommend. At the moment nuclear plant provides about 9 per cent of total gross electricity production throughout the Community.

Before going on to examine Community energy according to its various energy sources or sectors, it is useful to delineate how this energy is roughly apportioned. First, around three quarters of energy is consumed by industry and the domestic sector together. It is striking that such is the increased expectations for consumers that industrial and domestic consumption are about equal. The third largest consuming sector is that of transport, which absorbs around 15 per cent, leaving around 10 per cent of primary energy for conversion into secondary sources such as thermal electricity and manufactured town gas.

3.3 EEC ENERGY SECTORS

It is an interesting comment on the current state of industrialism in the Community that the total per capita energy consumption for each member country shows very little variation from the EEC average. However, it is in the sources of their supply that the individual countries demonstrate a marked diversity with, in consequence, profound and deepseated differences in their approach to the previously outlined attempts to draw up a common energy policy. A very brief summary of these differences is in order before embarking on a more detailed examination of the Community-wide energy sectors by turn.

Speaking in very general terms, oil is by far the most important source of the Community's energy and likely to remain so for the foreseeable future. It is also the most important source for each individual country except Luxembourg where coal is still king but which is so small as not to affect the Community percentages significantly. In Denmark and Italy oil totally dominates the energy economy.

Coal is a fairly clear second to oil in the Community as a whole although natural gas gained in percentage terms very recently. Coal is the second most important energy source for each country in the Community except for three members; the Netherlands and Italy, where natural gas is the second most important fuel and Ireland where peat still rates as the second energy source.

Natural gas is a close third to coal overall in its importance to the Community. It is an especially important energy source to the Benelux countries and Britain.

Thus, the Community may accurately be described as an oil dominated economy with coal, and increasingly natural gas, making up the bulk of the remaining energy required.

Each of West Germany, France, and Belgium conform closely
to the overall pattern; Italy, Denmark and Ireland rely especially heavily on oil, while Britain with oil, natural gas and
coal, and the Netherlands with natural gas, are special cases
whose ultimate contribution to the Community's energy larder
cannot be underestimated.

The chief worry of the Community in practical terms,
as distinct from abortive attempts at energy policy harmonisa-
tion, is its continued reliance on energy imports. Prior to
1973, the Community imported something over 60 per cent
of its energy requirements; both Britain and the Netherlands
managed to meet around 50 per cent of their requirements
domestically but most of the others needed to import around
75 per cent of their consumption. Following the measures
adopted in the wake of the OPEC oil price rise both by
the industrialised countries as a whole and by the Community,
but mostly by the responses of individual national governments,
the Nine had by 1976 reduced its energy import dependence
to around 55 per cent. While Britain's North Sea oil and
gas resources were likely to enhance the trend toward greater
Community self-sufficiency, the improvement in the Community economy was also likely to restore the continued
upward spiral in total Community energy consumption to
which the nuclear power plants could not foreseeably respond
significantly until the latter half of the 1980s. At this point,
having sketched in the overall contribution of the various
energy sources to the Community in very summary fashion,
it is necessary to examine the chief energy sectors in greater
detail.

3.4 OIL IN THE COMMUNITY

As we have seen in earlier chapters, oil has, since the early
1960s, come to dominate the world's energy economy much
as coal had done for several decades before. However, although
around 45 per cent of the world's energy requirements are
met by oil, the European Economic Community has, in terms
of its own consumption, relatively limited domestic oil
resources. This is an about-turn from coal which the Community possessed in relative abundance but which it allowed

to run down in terms of its coal producing capacity in the days when oil was both cheap and plentiful — roughly the period of the 1960s. The extent of this plentitude and the reasons behind it possibly need some explanation.

During the late-1950s and the early-1960s oil was in what some now regard as over-supply, with the major oil fields of the Middle East and North Africa being discovered and exploited under generally favourable commercial conditions. The not very surprising result was that oil prices fell relative to other costs with oil not only meeting most of the growth in energy demand but also making large inroads into traditional coal markets. By the early-1960s oil had begun to fulfil half the Community's energy requirements, in the process displacing coal as the principal fuel. However, by the late-1960s oil had begun to move into a position of relative shortage, more especially as the United States began overnight to become a major importer. The energy supply problem was aggravated for the Community by the fact that oil had proved such a potent competitor that the Community's coal industry had declined far faster than anticipated. In addition, the nuclear power programmes proved to be far slower in overcoming the technical problems of development than had at first been anticipated by the scientists, planners and politicians. The demand for oil therefore became that much more intense and with it an oil shortage began remorselessly to appear on the horizon.

Meanwhile, a new political force had arisen in the shape of the most powerful cartel that the world has ever known, the Organisation of Petroleum Exporting Countries, commonly known as OPEC. Its members comprised Abu Dhabi, Algeria, Indonesia, Iran, Iraq, Kuwait, Libya, Nigeria, Qatar, Saudi Arabia and Venezuela. Although established as early as 1960, it was not until 1969 with the vastly changed situation which had arisen between the producer and consumer countries that OPEC's membership effectively began to wield power on behalf of its member states.

The reasons for OPEC's enormous power worldwide and its central significance for Western Europe's oil supply in particular are not hard to ascertain. First, its membership controls over half of world production, or 80 per cent of all oil production if you exclude the Soviet Union and the

United States which are largely taken up with supplying their own requirements; second, its member countries harbour very nearly three quarters of the world's currently known reserves; and third, and most significantly for the Community's future, Western Europe obtains almost 90 per cent of its oil from OPEC members (even in 1974 this was still true).

The nature of the demands made by OPEC on the oil companies and their clients have been twofold: first, for higher prices and taxes and second, for greater control over operations in member countries. The results make a statistically dramatic picture. Between 1970 and mid-1973 oil prices rose by around 70 per cent, while in the few months following October 1973 prices more than quadrupled. As a concrete example, Saudi Arabian crude oil, which early in 1973 cost eleven dollars and seventeen cents per tonne, by 1974 cost fifty-one dollars and forty-seven cents per tonne. These increases put thousands of millions of dollars on the oil import bill for all the major Community member states. In addition, the Arab oil producers at one point cut back production by up to 20 per cent but this was circumnavigated by the oil companies re-routing supplies, leaving price as the chief continuing oil problem. As long as Saudi Arabia and Kuwait in particular are able and willing to carefully regulate the pressure — the Saudis held back an anticipated price rise in 1978 for instance against the odds — the cartel will remain in a strong position. At the latest, during the late-1980s the development of alternative energy sources will almost certainly curb the cartel's effectiveness and ultimately cause it to break down as a determining force. In the meantime Western Europe's energy import dependence is such that she must of necessity carefully monitor the possible courses of action open to OPEC even while she is gradually reducing her dependence upon Middle East sources of supply.

The pace at which the Community is capable of moving toward energy self-sufficiency is a matter of some controversy. Since the rate at which nuclear powered energy can be introduced is subject to severe restraints and continued uncertainties, the degree to which the Community can increase its oil self-sufficiency in the short and medium-term future is of considerable importance. As late as 1972 the Community produced a mere 12 million tonnes of its own domestic oil while consuming very nearly 800 million tonnes. It was never-theless already known that the North Sea was one of the world's major petroliferous basins. Thus the prospect of British entry into the Community has always held promise of a substantial movement toward oil and gas self-sufficiency. There are still considerable unknowns but to counterbalance the British expectation that they are capable of providing up to 90 per cent of the Community's domestically produced oil supply, there is a somewhat sobering estimate that only around a third of the Community's oil requirements is likely to be extractable from the North Sea whose oil is generally light, yielding less heavy fuel than other crudes and likely to cost between ten and twenty times as much as Middle Eastern oil to produce. The most obvious advantage of North Sea oil is that it is safe from direct external political pressures, close to the main centres of energy demand and thus prospec-tively, for the foreseeable future at least, still if barely competitive in European markets on the long haul.

The immediate prospects for North Sea oil are for a dramatic expansion. In the British sector, which accounts for a preponder-ant proportion of the total oil in prospect, oil production rose dramatically from 1.5 million tons in 1975, to 18.5 million tons in 1976, to 40 million tons in 1977 or 800,000 barrels a day. Since total Community energy consumption in 1976 was around 900 million tons of coal equivalent of which oil consumption comprised around 500 million tonnes, the latter figure likely to remain temporarily fairly constant, it can be seen that in 1977 the British North Sea fields were producing around one twelfth of total Community oil consumption. Most of this goes toward British direct requirements which absorb around 20 per cent of total Community oil consumption, but the figures nevertheless reveal the substantial and dramatic movement toward greater Community self-sufficiency that Bri-tain's indigenous oil clearly represents.

3.5 NATURAL GAS IN THE COMMUNITY

Although coal remains overall the second most important fuel in the Community, both because natural gas is often found together with oil and is in consequence in the northern part of the North Sea prospected by oil companies, and because its demand, especially in the domestic sector, is steadily increasing, it seemed logical to examine it ahead of coal.

Apart from synthetic natural gas (SNG) imported mainly from Algeria, there is no major international trade in gas as far as Western Europe is concerned though the potential deals between Algeria and both the United States and the Soviet Union are not insignificant; but again this is gas in a liquefied form. This means that the 17 per cent of total energy consumption within the Community which was met by natural gas in 1976 was exclusively from domestic sources of supply with the major exception of Norway which in the same year began exporting gas from its Ekofisk fields in the North Sea. This means that up until 1976 natural gas has made a major contribution to the self-sufficiency of the Community in energy by reducing its import dependence in one of the fastest growing fuels. Norway, with its close physical, political and economic relations with the Community does not really violate the principles of greater self-sufficiency which the Community is seeking to foster.

The two principal Community producers of natural gas are the Netherlands, which produce around half of indigenous natural gas, and Britain which produces around one quarter. The remaining 25 per cent came from Germany, Italy and France, in that order. Gas was first found almost two decades ago in 1959 in Groningen in the Netherlands. Since then the North Sea has yielded up several important gas fields. Although the expectation is that by 1980 the off-shore fields will be producing more than those on-shore — in the Netherlands, Germany, Italy and France — the bulk of existing reserves lie onshore in the Netherlands. The latter source alone could by 1980 conceivably be meeting around 10 per cent of the Community's total energy requirements. With this in mind, and the evident growth in demand in the domestic sector which now accounts for about one third of all natural gas usage within the Community, it is not difficult to visualise earlier estimates of 20 per cent of Community energy requirements by 1980 being met by natural gas proving to be underestimates.

There are not unnaturally a few hiccups in the pipeline such as the observable switch in 1975, particularly in Germany, to use coal rather than gas for power stations; and major questions on the horizon, such as the pricing policy of the British Gas Corporation which, by employing its monopoly purchasing rights to buy natural gas at half the price obtaining in the rest of Western Europe, seems at times to have slackened the rate of exploration in the British sector of the North Sea. Both of these issues will be examined in detail later.

But despite these qualifications, which are more or less inevitable riders to place on the record in 'guestimates' about energy mixes over so diverse a political economy as the Community represents, there is still a general belief that the British sector of the North Sea alone could be supplying enough gas to meet the existing total energy needs of Italy by 1980. By contrast, the on-shore gas reserves in France, West Germany and Italy are not very substantial and unlikely to play any more than a declining part in contributing to the Community's requirements for natural gas.

Apart from the North Sea and the possibility of more realistic pricing policies accelerating exploration in the British sector, there are as yet unrealised possibilities in the Celtic Sea for Britain and Ireland, the Western approaches for both Britain and France, and the Mediterranean for France. Although the Community's natural gas deposits are proving much more plentiful than when exploration first began on a serious scale in the 1960s, it must be remembered that, like its oil deposits, its reserves could be depleted at too fast a rate to be effectively substituted by the newer alternative energies including the various coal conversion processes. As will be made clear later in this chapter, the uncertainties that remain a characteristic feature of the various national nuclear power programmes make it imperative that no firm assumptions are made about natural gas meeting a major proportion of Community energy requirements in any but the long-term future.

3.6 COAL IN THE COMMUNITY

As we have already observed in an earlier chapter, in which we traced the attempts to forge an EEC Common Energy Policy, coal has registered a rapid decline from almost total dominance of the Community's energy economy. As late as 1964 in the original Six, and as recently as 1971 in Britain, coal was the chief source of energy. Coal production has in fact been falling fairly steadily at around 7 per cent per annum since the late-1960s. The principal causes for this decline were the greater competitiveness of oil and rising costs of both labour and materials. Since the recent spectacular surge in oil prices, the first of these causes has ceased to be valid and coal has theoretically become once again competitive with oil.

The problem for the Community has been that the coal industry generally had been running down for so long that it has been difficult to resuscitate rapidly enough to take advantage of the new situation. There are nevertheless a number of factors working in favour of coal in the medium and long term if not the immediate future, which are described later in this book. As far as the Community's coal industry is concerned, its greatest asset is its very extensive reserves, something in the order of 100,000 million tonnes at least which is calculated to sustain present energy needs for about seventy-five years. Britain is possessed of a high proportion of Western Europe's coal reserves; and among the most recent discoveries have been major new fields in Yorkshire and North Oxfordshire.

Coal has not only declined in the Community overall but also in all member countries. This has meant that national governments have been chiefly preoccupied with regulating the rate of rundown. For some time American coal was able to undercut indigenous European coal, a remarkable state of affairs considering the size of Europe's coal reserves and the skilled labour which then still existed. Britain, with Western Europe's largest coal industry, offers a partial explanation with its chequered labour history, though the principal cause seems to have been a chronic shortage of long term investment capital.

In any event, by 1972 the British Government was obliged to write off £400 million of the nationalised coal industry's debts. Whether in Britain, or in any other member of the Community, the coal industry was likely to attract the scale of investment necessary to generate a major revival in coal's contribution to the Community's overall energy supply remains

an important unanswered question. As a crucial supplement
to the question of how to attract adequate long-term investment
is whether, in Britain's case, a gradual denationalisation of
the coal industry might not in fact enable the degree of
integration with other energy sources to achieve that end.
Both these questions, concerning the means of obtaining investment and the practicality of some form of denationalisation,
are examined later in this book.

Meanwhile in 1975 coal production dropped in Germany
and Belgium, remained stationary in France and by year's
end had seen the closure of the last pit in the Netherlands.
Only Britain registered an increase due almost entirely to
the fact that in the previous year, 1974, there had been
a major coal strike. The British improvement thus somewhat
distorted the true picture and bestowed a legacy of a 6 per
cent increase in Community coal production in 1975 even
though, Britain apart, the major coal producing states in the
Community each recorded a decline. As the Commission noted,
in terms of output 1975 represented stagnation in the coal
industries of all member countries. The Commission anticipated
little change in the near future though there might be small
increases in output in Britain and France. Coal imports from
outside the Community were also expected to register a decline.

On the consumption side the picture was even bleaker.
In 1975 coal consumption within the Community fell by
9 per cent leading to stocks building up and with them
the burden of increased costs. The fall in consumption derived
from two principal causes: first, a fierce recession in the
steel industry (the worst for thirty years); and second, a
decline in the use of coal for electricity where it was to
some extent displaced by nuclear capacity but much more
by a decline in the demand for electricity by industry as
a consequence of the economic recession.

Turning from the recent past performance of the Community's coal industry (which may appear worse than it
really is by virtue of the severe impact on industrial energy
consumption of the economic recession which in much of
the Community may be on the mend) to the longer-term
factors, we are confronted with the inadequacy of investment
by comparison with any one of oil, gas or nuclear power
industries. Although investment in Community coalfields in

1975, chiefly in the Ruhr, Lorraine, Yorkshire and Midlands coalfields, estimated at £250 million, was 60 per cent higher than the previous year and rising to about £270 million in 1976, it palls into insignificance by comparison with the scale of investment being committed to the other energy sectors. As we shall see clearly in a later chapter when analysing the claims of coal for fair treatment in the future balance of the British energy economy, worldwide coal has a very promising future, enough to justify a more serious examination of how the coal industry in the Community can be 'reconstructed' to enable it to contribute once again an increasing share of the Community's energy requirements.

3.7 NUCLEAR POWER IN THE COMMUNITY

As outlined in the section of chapter two devoted to the EEC's Common Energy Policy, nuclear energy is anticipated to play an increasingly important role in meeting the Community's energy requirements and in reducing its reliance on imported sources of energy in the future. For the moment, however, nuclear energy represents a very small proportion of the Community's total energy supply, not much more than 1 per cent of primary energy consumption. But since the Community has already established itself as one of the world's major centres of nuclear research, ranking roughly third overall in importance to the United States and the Soviet Union, a massive investment has already been made in laying the basis for a nuclear power industry for the future.

Through the Euratom agreement the Six founder members of the Community began a nuclear research programme, with independent national programmes and American reactors either bought or built under licence. Today among the Community member states only Luxembourg, Ireland and Denmark are without nuclear power stations, although the last is hotly debating going nuclear. Among the Community's members, Britain's nuclear programme is arguably the furthest advanced, but it has recently run into substantial difficulties. In the process it has begun to reveal to the general public for the first time the combination of enormous long-term capital costs, the sheer accumulated risks, physical and financial, that it represents.

3.8 ALTERNATIVE ENERGY IN THE COMMUNITY

Short of a dramatic breakthrough in some particular sector such as solar energy, it is clear from the foregoing account that the Community will rely upon oil, gas, coal and nuclear power for the vast proportion of its energy supplies. Though their present contribution remains extremely marginal and it would be quite premature to place unqualified faith in their respective future contribution, it is more than likely that from their midst will be derived an energy source of major future importance. That among the ranks of hydro-electricity, geothermal power, tidal and solar power there is a future major source there cannot be much doubt; a great deal depends upon the amount of research which is devoted to their development. Such research must take into account not only the costs of generating the particular fuel in question, but the costs of transmission and storage which have sometimes been underestimated by proponents of a newly developing energy source.

Among the alternative energy sources in the Community, hydro-electricity is possibly the best developed and at the same time with the least potential for further exploitation. At the moment concentrated largely in Italy and France it provides around 3 per cent of total Community energy requirements. The absence of suitable sites for further development is an almost insuperable barrier for extending its contribution. Geothermal energy is confined to Italy and unless major new discoveries are made there is no foreseeable scope for development in the Community.

Tidal power has so far only been introduced in France where a barrage was built on the river Rance estuary in Brittany to take advantage of the extreme tides which characterise that part of western France. Although roughly comparable sites exist in other parts of Brittany, as in the western estuaries of England, the high cost of both capital investment per kilowatt of energy produced has effectively provided a disincentive for further exploitation, for the moment at least.

Of the four main alternative energies available in countries in the Community, solar energy probably offers by far the most promising prospects. Although considerable experimentation and research is being carried out in several member countries the prospect is that the comparatively more extensive programmes being sponsored in both Japan, and especially the United States, may offer the earliest possibility of introducing solar energy as a major contributor to Community energy, especially domestic energy supplies.

3.9 SUMMARY OF COMMUNITY ENERGY

Since the preceding sections constitute summaries in themselves, some general observations on the overall structure of energy supply within the Community is probably in order. The first and fairly obvious observation is that in most member countries the majority of energy industries (especially coal and electricity) are under direct state control. In spite of that fact the degree of co-operation between various state industries has been very small. The second point is that in the case of oil, still the principal energy resource, despite Britain, France and Italy having major state interests in oil companies, despite the British National Oil Corporation, despite the British and Dutch participation in Shell (one of the seven major world oil companies), a major degree of control of the Community's energy supply rests with American multinational oil companies. While these same US companies (constituting more than half of the activity of oil companies operating within the Community) are now beginning to operate under much more stringent conditions in the United States as well as elsewhere, the balance of both preserving and maximising the benefits to the host country, and at the same time providing sufficient incentives to attract and retain such companies, is among the central issues that the Community member states must face in the next few years.

It is because British energy policy — and more especially the strategy that the British Government adopts in handling what is not only its own greatest natural resource but also, as we have demonstrated in this chapter, among the most vital resources to be disposed of within the Community — is so crucial to the Community's members as a whole that this book will devote so much space to British energy policy.

The following table shows estimated primary energy consumption for the period 1975 to 1985, assuming that no changes are made to existing energy policy and based on an expected average annual growth figure for the whole economy of 4 per cent:

TABLE 3. *Primary UK Energy Consumption, 1975–85*

	1975	1980	1985
	mtce (%)		
Fuel oil	181 (52.1)	216 (50)	226 (45)
Hard coal	66.5 (19.1)	72 (17)	73 (15)
Brown coal	34.4 (9.9)	35 (8)	35 (7)
Natural gas	48.7 (14.0)	73 (17)	87 (18)
Nuclear energy	7.1 (2.0)	28 (6)	62 (13)
Other	10 (2.9)	11 (2)	13 (2)
	347.7	435	496

PART TWO

THE PATTERN OF FUTURE POLICIES

4 German Energy Policy

Not only is West Germany indisputably the most powerful and prosperous country in the European Economic Community, but after Britain the effects of its national energy policy are likely to be the most pervasive of any other partner in the Community. This is principally because its energy consumption, arising from its large industrial base, its population and its extremely high levels of domestic consumption, is by far the highest in the Community at very nearly thirty per cent of total consumption. Together, Britain and West Germany dominate the Community market for domestic energy consumption. In 1973, for instance, the two nations accounted for considerably more than 50 per cent of total Community energy consumption comprising 46 per cent of all oil consumed, a roughly similiar figure for natural gas and a striking 70 per cent of total coal consumption. It is therefore a logical step to devote a separate chapter to such a major element in the political economy of the European Community's energy market. For despite the rate of increase in consumption having been slowed by the OPEC oil price rise, West German consumption is expected to reach 475 million tons of coal equivalent by 1980 and 555 *mtce* by 1985.

4.1 GERMAN ENERGY INDUSTRY

As a result of the OPEC embargo and price rises, more

than quadrupling the price of Germany's imported oil, the
Federal government was obliged to revise completely its
national energy strategy. Moreover, despite a great variety
of efforts to ease the pressure on the previously existing energy
economy of West Germany, it was very soon apparent that
Germany would be dependent on imported oil until the beginning of the 1980s at the very earliest. In response to this
new measure of uncertainty in the energy economy, involving
the politicisation of what had previously been an essentially
commercially determined environment, certain features of the
German energy industry need enumerating.

(1) Although West German energy sources are compara-
     tively diverse, she is nevertheless dependent on imports
     for around 57 per cent of her total requirements.

(2) West Germany possesses considerable deposits of both
     hard coal and lignite, the latter constituting the largest
     source of lignite production in Western Europe.

(3) Although it is among the most advanced nations any-
     where in nuclear energy, West Germany only obtains
     around 1 per cent of its primary energy from nuclear
     sources so far.

(4) Half of West Germany's natural gas supply is derived
     from domestic sources which, certainly inshore, appear
     strictly limited. This means that if natural gas is to
     play an increasing part in meeting West German energy
     requirements, it will almost certainly need to come
     from external sources; whether principally from outside
     the Community or from within it has yet to be deter-
     mined.

(5) With oil meeting 55 per cent of primary energy consump-
     tion in West Germany in 1973, oil from domestic sources
     meets only around 2.5 per cent of total primary energy
     consumed.

(6) Thus, by way of summary, the very approximate division
     of West Germany's energy consumption according to
     source is 50 per cent of total energy consumption is
     oil, 20 per cent gas, 20 per cent coal, which on the
     basis of the earlier five points only serves to underline
     the continued dependence of West Germany on imported
     energy, either of oil or gas, or even coal.

The important question, both for the future stability and security of supply of West Germany, and the general prosperity and security of the Community as a whole, is to determine the nature and extent of West German energy import requirements and the extent to which they can be met from within the Community either by the Netherlands or Britain. The answer to these central questions lies to a large extent in the analysis of chapter 5 insofar as Britain has a part to play, also in the conclusions and proposals of the final chapter. However, some guidance as to West German energy import demand can be given here in a very summary and approximate form.

The demand for natural gas imports is currently substantial and already constitutes around 10 per cent of total primary energy consumption. The demand for oil imports is even greater still, accounting for around 45 per cent of total primary energy consumption in West Germany. Thus, in the very area where West Germany will need guarantees of future secure sources of supply of both oil and gas, another partner in the Community, namely Britain, has the potential for greatly increased domestic production. How much Britain is able to make available for export, or at the very least sharing in an emergency, depends heavily on the resourcefulness and flexibility she demonstrates in the exploitation of her principal energy sources, and the resolution and restraint she is able to engender among her citizens in pursuit of the much more rigid conservation of their indigenous energy resources.

4.2 GERMAN ENERGY POLICY OBJECTIVES

Having described the general structure of the German energy industry in terms of its dispersion of various energy sources and the degree of its import dependence, it is superfluous to belabour the fact of the unpredictability of energy supply both from OPEC countries and elsewhere. Overall German national economic policy has to incorporate a coherent set of principles or objectives to deal with the expected continuing fluctuations in energy supply and demand. These principles, enunciated by the West German Federal Ministry of Economics, include the following six objectives:

(1) higher priority for securing oil supplies and reducing its share in the total energy supply;
(2) accelerated use of nuclear energy, natural gas and lignite;
(3) a new position for hard coal;
(4) greater saving on energy;
(5) higher priority for energy research and
(6) improved contingency planning.

Then, as a practical follow-up to the post-1973 energy priorities there is a reassertion of the ongoing aims of German energy policy, as formulated prior to 1973. These are summed up in four basic aims:

(1) sufficient low-price energy to meet consumer demand;
(2) securing supply in the medium and long term;
(3) providing energy at favourable total costs to the economy, on a long-term evaluation;
(4) adequate and early consideration of the requirements of environmental protection in order to bring the often conflicting demands of energy requirements of the current industrial economy into line with minimum environmental considerations.

4.3 GERMAN ENERGY, 1975–1985

In the light of the foregoing analysis a precise energy policy would be out of place. However, the Federal Government favours a qualified forecast of the medium-term results it expects from the pursuit of certain emphases whose chief value is that it may offer certain guideposts for government officials and also for industrial executives and management. But since the uncertainties in the behaviour of the international energy market remain, the following quantitative framework must be susceptible to rapid adaptation to changed levels of demand and supply lying outside the previous patterns. The initial question that the West German energy planners asked was, what share domestic sources of energy can and ought to contribute in view of the changed situation. The second question was what contribution could be made by comparatively secure forms of energy, such as natural gas and nuclear power. In asking both these questions, it will become apparent that, at least up until the mid-eighties, a very considerable proportion of energy demand will still need to be met in West Germany by oil.

There are a number of important points about West German energy allocation possibilities in the medium-term future, and the recent past for that matter. The most important aspect is the apparent decline from a position of overdependence on imported oil. Since this is arguably the chief post-1973 objective of almost every Western industrial economy, this trend is not to be despised. On the other hand, its achievement should not be overrated, with the reduction from a dependence on imported oil as a percentage of total energy supplies dropping from 55 per cent in 1973 to 47 per cent in 1980 and about 44 per cent in 1985. This still leaves oil as the most important single source of energy for West Germany in the mid-1980s.

By itself, this is a substantial reason why Britain should do all in its power to help meet German requirements, given the complementary character of not only British and German requirements but also the interrelated political economy of Western Europe and the aim of the European Economic Community to create the greatest possible degree of self-sufficiency in energy supply. The table is also a reminder of the time it takes to develop some of the so-called secure sources of domestic energy. A further conclusion would be that both natural gas and nuclear energy are likely to record fairly high growth rates. Together with lignite, they will reach a share in primary energy supply of roughly 40 per cent in the mid-1980s compared with only about 33 per cent in the earlier 1973 forecast. The maintenance by coal of its present position in West Germany is also an interesting consequence of the OPEC price increases for oil.

The previous figures, as with any estimate, are subject to a great many uncertainties about the development of world energy markets, about the rate of progress in either the exploration or scientific development of individual forms of energy, etc. Beyond 1985 those uncertainties increase as the scope broadens to embrace factors well beyond the energy economy itself, such as the development of the national economy as a whole, the direction taken by the world economy and the influence of cultural factors which may create entirely different consumer appetites. Beyond the mid-1980s there is no estimating what new forms of energy current research and development programmes may have by then uncovered, either in the form of alternative forms of energy or merely more sophisticated energy technologies which refine the efficiency of current energy sources. The quantifiable energy framework can never be wholly relied upon, even as a set of starting points, beyond the medium term, in this case the mid-eighties.

4.4 GERMAN AND EUROPEAN ENERGY

In its awareness of the interdependence of the world energy market Germany is second to none, not least because of its continuing dependence on imported energy whose supply in terms of both price and quantity remains unstable. In order to respond to this situation positively the German Government has recommended that international co-operation in the sphere of energy policy should concentrate on the following aspects

(1) The industrialised countries must find common means of securing supplies which eliminates wasteful competition between themselves.

(2) The industrialised countries must make a concerted attempt to reduce energy dependence on oil by diversification into other energy sources.

(3) The Industrialised countries must establish a new means of 'equilibrium' with the oil producing countries.

(4) In the light of the escalation of energy demand amongst virtually all industrialised countries, there exists an absolute necessity to collaborate on an international level if new sources and new technologies are to be developed.

(5) Bilateral arrangements between producer and consumer countries must be supplemented and complemented by multilateral arrangements if some form of durable balance is to be struck within the wider international framework.

In pursuing these five objectives the role of such bodies as the International Energy Agency will be crucial.

Within the framework of the European Economic Community where, as we have seen, progress has been so slow, the German Government has indicated its own priorities: strengthening the security of supply (particularly by making better use of energy available within the Community), co-ordinated efforts to obtain additional energy and, not least, comon research efforts in new energy forms and improved technology in the utilisation of the conventional energy sources.

In the sphere of international and European energy collaboration the German Government's energy policy embraces improved contingency planning of which the most notable feature is an ambitious stockpiling policy for mineral oil, including crude oil stored in underground caverns for security reasons, constituting a ninety day supply. The German Government has also decided it is wise to build up hard coal reserves from domestic production of the order of 10 million tonnes with the mining companies. The advantage of such a coal reserve is to be seen not only in times of oil import uncertainties when coal, for instance, can be substituted for heating oil, but also to cope with prospective delays in the completion or breakdown in the construction and generation of nuclear power plants.

4.5 GERMAN OIL SUPPLY

Among the most important trading countries in the world, whose industrial output together with that of the United States and Japan accounts for half of the total output of the Western industrial countries, Germany is also one of the major importers of oil. As such it has a vital interest in a free and properly functioning international oil market. Since around 1970 the balance of forces in this market has altered drastically. The international oil companies, for instance, find themselves in a position where their traditional concession rights are being reduced and their control over crude oil is lessened. By virtue of their extensive processing and marketing capacities and their technical and economic potential, the oil companies, though substantially circumscribed compared with formerly, will continue to play an indispensable role in the world energy market. During the oil crisis of 1973–4 and the dramatic expansion of oil company profits in the wake of the OPEC price rises, these companies came under increasing criticism in consumer countries. The German Government's response to the criticism of the international oil companies has been to support the drive for greater transparency in their marketing activities. Such a policy of disclosure, argues the Federal government, would not only satisfy those who may be concerned that the national interest is being subverted but also serve to promote the oil companies' best interests by revealing its own impartiality among consumers.

Meanwhile, as the power and influence of the energy multinationals have relatively declined, or apparently so, the control of national oil companies in the producer countries over the production of crude has grown rapidly and with it their influence in the international oil market as a whole. This accretion of power, however, does not apply to national oil companies in the consumer countries. Here companies which were previously supplied by the international companies have been obliged to switch to other sources. Their position in the market has been further impaired by the differences between the price of equity oil and the free world market price. Yet the contribution that these companies make to supplying Germany remains essential. In favouring a policy of non-interference in the German oil market, the Federal Government implicitly regards this as the best means of a healthily structured market; it is also on record as favouring a large number of efficient, small and medium-scale businesses in the energy market as elsewhere.

One of the distinctive features of German energy policy since the 1973 crisis has been the stepped up effort by the Federal Government to implement the 1973 Energy Programme objective of creating a strong German mineral oil group which can join in international co-operation, especially with the oil-producing countries, as an equal partner. By purchasing a majority share participation in Gelsenberg Ag it has initiated the process of reorganisation. The activities of Veba and Gelsenberg are being brought together under the common management of Veba Ag.

The rationale of creating a German national oil company arises out of the Federal Government's efforts to broaden the country's own crude oil base in the wake of the 1973 crisis. As the largest consumer of oil in Europe, Germany has a particular interest in joining in the worldwide search for oil. The chosen instrument for the Federal West German Government in this field is called Deminex, an organisation already widely accepted by the international oil companies as an equal partner. Starting virtually from scratch, Deminex is carrying out test drilling in the British North Sea, Nigeria, North Africa, the Caribbean, Peru, Canada, the Middle and Far East; from 1975 to 1978 the German government has earmarked 800 million D-mark for investment in Deminex.

Finally, in terms of refinery capacity, the German Government attaches great importance to a regionally balanced refinery structure, that is, one which is geographically dispersed for both socio-economic and defence reasons. The opportunities for securing long-term supply arrangements with oil-producing countries are to be taken whenever they arise.

4.6 GERMAN NATURAL GAS SUPPLY

After mineral oil and hard coal, natural gas is the third largest source of primary energy with a share of a little more than 10 per cent of total energy supplies. This is a significant drop from the still dominant oil, which still meets more than 50 per cent of Germany's energy requirements as we detailed earlier in this chapter. The German natural gas market has been, with some fluctuations from year to year, met about equally by foreign and domestic sources. Among the most politically significant developments since 1973 in Germany's natural gas market has been the addition of the Soviet Union to the Netherlands as one of the two major foreign suppliers, followed in 1976 by Norwegian natural gas from the Ekofisk field. The expectation is that by 1980 Soviet natural gas supplies to West Germany will have reached the 10 billion cubic metres a year level, while the Norwegian field will be supplying around 6 billion cubic metres annually via the subsea pipeline to Emden. Meanwhile, negotiations are under way with Algeria for the delivery of a total of

12.5 billion cubic metres annually, also with Iran with whom there is a major effort to create a broadly based framework of co-operation in the economic sphere. In the case of Iranian natural gas the plan is twofold: by land pipeline in collaboration with the Soviet Union, and by sea.

It is now a conscious and declared aim of German energy policy to promote an increase in the share which natural gas contributes to West Germany's overall energy supply. As first announced in the 1973 Energy Programme, the Federal Government will actively support companies in the securing of new supply quantities for the German market in the following manner: first, by general policy measures, second, by international bilateral agreements and thirdly, by public financial guarantees. The Federal Government hopes that by 1980 the pursuit of this strategy will have raised supplies of natural gas on the German market to around 87 million tons of coal equivalent.

As part of the overall campaign to raise the proportion of natural gas, the Federal Government has already set aside 40 million D-mark for deep drilling for natural gas within the borders of the Federal Republic, chiefly in Miesbach, Bavaria and Velpke-Asse in Lower Saxony. The German section of the North Sea has also been resumed as an area of sustained exploration for natural gas. In closing this brief resumé of German natural gas supplies it needs emphasising that cross-frontier co-operation between major gas companies is becoming an increasingly common feature of the development of natural gas in Western Europe. Apart from the scale of capital investment required, the need to share transportation costs is an obvious requirement if natural gas is to be expanded both in absolute terms and in terms of its proportion of the total energy supply in the future.

4.7 GERMAN COAL SUPPLY

Ever since 1973, when the importance of Germany's coal reserves became strikingly apparent, the Federal government has sought to apply an optimum policy toward Germany's domestic coal reserves. The extent to which Germany lent on her coal stocks at the height of the Middle East crisis of 1973-4 can be seen in the decline on coal stocks from 19 million tonnes in September 1973, to somewhere below 5 million tonnes a year later. Much of this additional demand went to other European Community countries, so in practice the German coal companies provided a significant back-up function for the rest of the Community.

In its 1973 Energy Programme, released before the oil crisis broke, the German Government suggested that the German coal industry should aim at a target of 83 million tonnes in 1978. However, the Middle East crisis very soon presented the international coal market and ultimately the German coal industry with an entirely new situation. The German Government responded to this situation by calling upon the coal industry to make the fullest use of its current production capacity and to take the most careful assessment possible of the German market when deciding how much coal production should be exported. Since this immediate post-Middle East crisis period the world boom in steel has undergone a recession and demand for coal has levelled off considerably with the price of coking coal declining in consequence.

Under the new conditions, the general competitiveness of the German coal industry has become exposed as being less impressive than most of her general industrial structure. The two major coal producers who have traditionally exported to Western Europe — the United States and Poland — have signalled their intention of substantially increasing their respective coal outputs to which the German coal industry is not especially well equipped to withstand for a variety of reasons, principally the unfavourable geological conditions which have added to the wage costs of extraction. The possibility of a widening gap between coal earnings and coal costs lies ahead of German domestic hard coal production. The balance between the need to maintain a coal industry at home and the costs of so maintaining such a structure are a problem of some delicacy.

At the latest count the following markets for coal will exist in Germany in 1980, broken down into three major sectors, namely 35 million tonnes for the electricity industry, 25 million tonnes for the domestic iron and steel industry, 11 million tonnes for consumption in the home and the remainder of industry and 18 million tonnes for export (on the assumption that they become profitable to the coal exporting companies, not always the case in the 1973–4 period when Germany saw the industrial sustenance of the Community as a higher priority than immediate profitability). Planned coal reserves of 10 million tonnes and imports of around 3 million tonnes would give total coal sales of around 90 million tonnes. Such a figure, subject to a great deal of qualification due to the uncertainties of the international energy market as a whole, the unknown consequences of revised energy policies in the main consumer countries, as well as the details and impact of the International Energy Programme, is roughly equivalent to the planned capacity for natural gas. The fact that 80 per cent of German coal sold domestically goes to the electricity and iron and steel industry, highlights its central contribution to the German industrial infrastructure, more especially since it is government policy to make electric power supply as independent as possible of imported energy, most especially mineral oil.

Recently the German coal industry has contained an important export component, in 1974 reaching as high as 33 million tonnes. By 1980 the Federal Government calculates this will have levelled off at about 18 million tonnes, mostly coking coal for the furnaces of the iron and steel mills of other Community members. However, much will depend upon the eventual coal policy of the Common Energy Policy with the German Government anticipating a general decline in her coal exports for the foreseeable future. With a degree of uncertainty built into her coal industry, one of the chief problems is to preserve the skilled labour force by offering an attractive career for the future. At the moment foreign workers represent about 20 per cent of the underground workforce in German coal mines, a factor which needs to be carefully monitored within the context of tighter rules and safeguards for guest workers.

After hard coal and natural gas, lignite, or 'brown coal', comprises the most important source of domestic energy in West Germany. One of the cheapest forms of energy, more than 80 per cent of lignite output is consumed in electricity generation. Particularly auspicious as far as future German lignite supplies are concerned is the recent opening up by the companies concerned of the Hambacher Först. With the help of special depreciation allowances to encourage the opening up of reserves the expectation is that this will significantly augment Germany's coal supplies from domestic sources. Germany's lignite reserves are the only known major commercial source contributing to the Community's coal supply, other than imported lignite from the United States or Poland.

4.8 GERMAN NUCLEAR POWER

With a currently installed capacity of around 2,500 milowatts, Germany's nuclear power stations provide not much in excess of 4 per cent of electricity generated in West Germany. In the 1973 Energy Programme the Federal Government aimed at installing 18,000 milowatts by 1980 and 40,000 milowatts by 1985; since the Middle East crisis of 1973-4, the Federal Government has raised its targets to 20,000 milowatts by 1980 (ie around 25 per cent of electricity generation) to 45,000 milowatts by 1985 with a hoped-for attainment of 50,000 milowatts (ie around 45 per cent of electricity generated). In this respect Germany resembles France in her professed determination in the face of significant opposition to raise dramatically the capacity of nuclear powered generating stations in the medium-term future.

The development of two types of advanced nuclear reactors is being pursued in Germany: the high temperature reactors, because of their high thermo-dynamic efficiency, help render the cooling and waste heat problem less acute; the fast breeder reactors in the long term could considerably increase security of energy supply and perform a regulatory function on the uranium market. Whereas the first type of advanced reactor, the high temperature models, could be widely introduced within roughly a decade, the fast breeders are not expected to come into service commercially until sometime in the 1990s. At the moment a prototype plant based on a sodium-cooled fast breeder reactor (SNR-300) is currently being constructed at Kalkar with the participation of Belgium and the Netherlands.

The extremely high costs, not to mention economic risks, involved in developing these advanced types of reactor make it increasingly necessary to look in the future for opportunities for international co-operation and division of labour. With her pressing economic problems and her comparative abundance in other forms of energy, Britain, despite her early start in nuclear-powered generating plants, only recently took an interim decision on the type of reactors in which it wished to expand its investment, and is in consequence less well-placed than might have been anticipated a few years ago. Nonetheless, already in the area of uranium procuration there has been evidence of greater international co-operation with Germany obtaining her supplies of enriched uranium from the American Atomic Energy Commission, while URENCO, the German-British-Dutch centrifuge company is likely to grow in importance.

4.9 GERMAN ENERGY CONSERVATION

The concept of a concerted policy of energy conservation is of very recent derivation, at least in the postwar Western industrial countries. Energy everywhere has become both expensive and available in far from unlimited quantities. The minimum aim, therefore, of almost all national energy policies in the realm of conservation, including Germany's, is to slow the growth in energy consumption by making more rational use of the energy available. Even in the short term this has the effect of reducing the vulnerability of the economy to dislocations in supply. A more sustained benefit to a country like Germany, both highly industrialised and densely populated, is that it reduces or at the very least slows the advance of environmental pollution. Whether in fact the conservation measures do any more than slow the social and physical corrosion that appear an inevitable concomitant of our industrial society is a subject beyond the scope of this chapter.

Meanwhile the opportunity for saving energy is adjudged by the German Government as relatively large. In the first place a 20 per cent loss of energy occurs in energy production and its transfer to the consumer; in the second place, the loss at the consumer level is generally estimated to exceed 50 per cent overall. With current technology in use, only part of this loss can be immediately overcome, but it is this same part which can be avoided which German policy has sought to solve. These avoidable losses typically have three principal forms: insufficient heat insulation of buildings, insufficient heat regulation and poor servicing of installations, and possibly most wasteful of all, the almost total failure to re-use waste heat from industrial processes in particular. In each of these spheres, while government legislation and practical information is indispensable, the final responsibility is seen to lie with the consumer, whether the individual householder or the individual industrial consumer. As in other countries, Germany sees an important contribution to future conservation is likely to be made by bringing the scientific and business communities into closer collaboration on this subject, an approach which will be more fully explored in the concluding chapter.

The Federal Government has also supported a more rational use of energy through selective legislative measures in a variety of fields. As early as January 1974, a Federal decree suggested to the state governments and other regional authorities that they improve the thermal insulation of both domestic and public buildings. Since space heating absorbs the largest share of primary energy consumption, around 40 per cent, and energy losses are particularly high in this area, it was a particularly urgent development. New Federal legislation was also introduced for the mandatory introduction of better heat insulation into new buildings. The government has subsequently commissioned the preparation of an expert opinion on a waste recycling programme, more especially of energy-intensive industrial raw materials.

To maximise the effectiveness of energy conservation measures, the government initiated a Framework Programme on Energy Research, 1974–7 which embraced the following research projects:

(1) an analysis of the energy flow in West Germany and of the consumption of energy by households and industry with special attention paid to tax measures;

(2) studies of energy requirements and supply, availability of energy, energy economics, environmental pollution and adaptability of the existing supply structure;

(3) preparation of 'heat maps' in collaboration between industry and scientific research institutions as planning

TABLE 4. *Germany, OECD Energy Balances, 1977*
Energy Balance Sheet (MTOE)

	Solid Fuels	Crude Oil & NGL	Petroleum Products	Gas	Nuclear Power	Hydro & Geothrm	Electricity	Total	P.E.E.E.*
Germany 1975									
Indigenous Production	89.86	5.82		14.68	4.76	3.81		118.92	
Imports (+)	6.56	92.77	37.71	20.72			1.52	159.27	
Exports (−)	−15.92	−0.01	−6.44	−0.07			−0.84	−23.29	
Marine Bunkers (−)	—	—	−2.83				—	−2.83	
Stock Change (+ or −)	−7.70	−3.22	2.62	−0.29			—	−8.58	
Total Energy Requirements (TER)	72.80	95.35	31.06	35.04	4.76	3.81	0.67	243.49	
Statistical Difference	0.32	4.42	−3.93	−0.16			0.10	0.76	0.25
Electricity Generation	−42.19	—	−4.78	−11.60	−4.76	−3.81	25.95	−41.18	67.13
Gas Manufacture	−2.50	—	−0.44	2.76			—	−0.17	—
Refineries	—	−99.78	91.19	—			−0.38	−8.97	0.96
Own use by Energy Sector and Losses	−4.92	—	—	−0.79			−4.05	−9.76	10.22
Total Final Consumption (TFC)	23.52		113.11	25.25			22.29	184.17	56.19

82

~~Total Industry~~	~~15.95~~		~~26.96~~	~~16.18~~			~~11.33~~	~~72.42~~	~~28.58~~
of which									
Iron & Steel	11.92		3.17	6.18			1.73	23.00	4.37
Chemical	1.25		4.18	5.34			3.59	14.36	9.04
Petrochemical	—		5.95	—			—	5.95	—
Other Industry	2.77		15.66	4.97			6.01	29.41	15.14
Transportation	0.28		31.91	—			0.76	32.95	1.92
of which									
Road			28.01	—				28.01	
Rail	0.28		0.64				0.76	1.69	1.92
Air			2.35					2.35	
Navigation**	—		0.91					0.91	
Other Sectors	7.29		44.52	8.77			10.20	70.78	25.72
of which									
Agriculture	—		1.35	—			0.55		1.37
Commercial Use	—		—	—			—		—
Public Service	—		—	—			1.49		3.75
Residential	7.29		43.17	8.77			8.17		20.59
Non-Energy Uses	—		7.72					7.72	
(Not included elsewhere)									
Electricity Generated — GWH						21398	17110	301802	
Efficiency in Electricity Gen. — percent						38.7	38.7	38.7	

83

Stock Drawdown +/Stock Increase —
MTOE = Millions of Tons of Oil Equivalent

* — P.E.E.E. = Primary Energy Equivalent of Electricity
** — Internal and Coastal Navigation

documents for heat distribution systems and the use
of district heating for instance;

(4) comprehensive studies on 'technologies for the conservation of energy', with a view to establishing a survey
of the various possibilities of a more rational use of
energy.

It is in fact in the field of district heating that the Federal
West German Government attaches particular importance,
most especially long distance heating plant systems, since these
hold the potential not only of conserving energy on a substantial
scale but also greatly reducing the pollution of the environment.

Finally in the field of energy conservation, the Federal
German Government supports the development of technologies
which will lead to the conservation of energy in a variety
of sectors of the economy where energy consumption could
be greatly reduced without effecting output but by improving
conversion efficiency.

4.10 GERMAN ENERGY INVESTMENT

As we noted in the chapter on energy policy in the EEC,
specifically in the section on coal in the community, section
3.6, even when the post-1973 rise in oil prices had taken
place giving coal a renewed theoretical advantage, the revival
of the respective national coal industries in Western Europe
failed to take place. One of the most obvious and persistent
reasons for this generally observed fact was the relative inability, for a wide variety of reasons, for the respective coal
industries to attract sufficiently large-scale investment capital.
In Britain, the largest national coal industry, this feature
has to be seen against the paucity of industrial investment
generally throughout most of the postwar period, yet the
problem is far from being confined to Britain.

It is apparent that the kind of priorities outlined in the
preceding pages on German energy policy demand an increasing commitment of *long-term* capital resources. Moreover, the
accelerated extension of secure energy resources as well as
the necessary transformation, transport and processing capacities for German energy supplies all point toward an increasingly

larger volume of investment. In constant prices the total investment requirements of the German energy industry is estimated
to be in the order of 250 billion D-mark up to 1985. This
is equivalent to about 21 billion D-mark annually. This figure
should be compared with the pre-1973 levels. The investment
figure for 1973, for instance was roughly 15 billion D-mark.

Speaking very broadly the Federal Government has studied
the implications for the capital market of the future additional
financing requirements of the energy industry and reached
the general conclusion that no special capital market terms
need be created. This observation embraces the electric power
industry which is notoriously difficult to finance in many
industrial countries. The only foreseeable major bottleneck
is the capitalisation of the electric power distribution companies
which are to build nuclear power stations. The cost per
unit of installed capacity in nuclear power stations is about
70 to 80 per cent above those of conventional power stations.
Moreover, the pure building time is about twice as long
as that required for conventional power stations of the same
size.

The financing requirements for nuclear power stations can
be met within the framework of internally generated funds,
above all by depreciation of the stock of power plants now
in service. Until larger sums become available after entry
into service of nuclear plants, a considerable gap will arise
between the growth of investments and the possible depreciation allowances. The balance will have to be met by additional
equity resources or borrowed capital. The crux of the problem
which has to be tackled by the German energy industry
is the financing of nuclear power stations during the transitional
period from the late-1970s to the early-1980s. So far the
Federal Government believes the industry does not require
any financial subvention.

5 British Energy Policy

It has been one of the sadder features of the postwar British economy to be more concerned with the distribution of income than with the utilisation of resources. To this energy policy is no exception. In their progressive accumulation of statutes and regulations affecting the energy market, successive British Governments have systematically eroded the characteristic qualities of free competition — the freedom of entry for producers and freedom of choice for consumers. Moreover, the non-observance and sometimes open breach by governments of statutes designed to preserve the independence of some of the state energy sectors has been scarcely less marked. In the light of this there need be no apology for devoting a considerable part of the earlier section of this chapter to the history of the exploitation of the North Sea to date, even though the current public debate has tended to focus almost exclusively on the undoubtedly crucial elements of energy policy for the post-1985 era.

Considering that oil and gas constitute around 60 per cent of Britain's current energy supplies, and are likely to do so until the late-1980s at the very least, this should surprise nobody. It is far from being too late to make a radical reappraisal of the means of extracting and taxing the oil and gas that lies not only in the North Sea but also in other places around Britain's coasts. Moreover, since much of the United States' major reserves of oil and gas lie offshore and thus open to closer regulation than the traditional inshore sites, the British experience has considerable relevance to future US domestic energy policy (as well as US energy companies operating in the Eastern Hemisphere).

This chapter falls naturally into three parts. The first part is a critique of the current framework for extracting Britain's principal currently available energy resource, North Sea oil and gas. The second part is a brief enunciation of six energy policy principles. While the third part is composed of proposals for some priorities to be followed now to meet the challenge of the post-1985 period in the realms of coal production, nuclear power, conservation and renewable resources.

Some preliminary points need making to demonstrate the central part that energy policy must inevitably play in the context of the operation of the British economy. As the January 1977 Energy Policy Review makes clear, the British energy industry employs 750,000 people (3 per cent of the working population); capital investment in 1976 was £3 billion (15 per cent of the country's gross fixed capital formation); annual turnover is over £11 billion; it produces 5 per cent of GDP and in 1975 accounted for 18 per cent of the total import bill. Discounting the last element, which by 1980 can be expected to be negligible and more than compensated for by energy exports, these figures underline the sheer size of the energy sector within the economy but they give no indication of the ramifications of energy policy as they effect the national economy in general and government policy in particular in the spheres of economic, industrial, social, regional, environmental and international policies.

5.1 INTRODUCTION: NORTH SEA OIL AND GAS

For some generations past Britain has imported almost all her raw materials and more than half her food. Today she faces an almost unique prospect. Alone among the major powers of Western Europe, even of the Western world, she has the capacity for self-sufficiency in energy by 1980. In this stark probability lies both the hope of industrial regeneration from the breathing space it offers, or if abused the haunting risk of chronic financial insolvency; if she can manage this great treasure house of natural resources with foresight and skill over the next decade it could provide among the most vital ingredients in her long sought economic and industrial regeneration. Equally, the very same resources badly handled could lull her people into a false sense of security leaving her at the end of the 1980s languishing without either energy or an adequate industrial infrastructure, like some great whale stranded on the beach.

The determining factor in the handling of these undreamt energy resources lies in the structures that are being created and investment decisions being made between now and the end of the decade. It is not that in principle decisions in the energy sector cannot be amended; merely that in practice the state sector has shown itself less amenable to responding to rapidly changing demand, circumscribed as it has been by a host of non-commercial operating criteria. Moreover, where government has such a growing direct day-to-day surveillance, if not involvement, in the entire or almost the entire energy industry, there is an inevitable loss of objectivity about the broad long-term policies which governments by their very constitutional nature are best suited to devote themselves. One need look no further than the government's increasingly abortive attempts to regulate prices and wages to see how counter-productive to both prosperity and equity (eg differentials) such detailed direct government management of what is rightly the sphere of the market to determine within certain social constraints. To clarify the meaning of such general principles and very broad assertions it may be useful to outline four factors which have recently pervaded the climate of the British energy sector.

(a) *Indebtedness*

That Britain's overall indebtedness arises from a long-term relative economic decline is rarely disputed. What is also fairly apparent is that Britain's colossal indebtedness — in which recently her reserves were only sufficient to pay the interest on her debts for the following four years — could not have reached this dangerous point without the prospect of massive earnings from her North Sea oil and gas fields. As at the end of 1976, the Department of Energy estimated that Britain's oil and gas reserves in the North Sea stood at something in the order of £300 billion, two thirds of which derived from oil reserves, a calculation based of necessity on current prices and exchange rates. By 1980, the Department claims, North Sea oil could improve Britain's balance of payments by over 3 per cent of gross national product and by 1985 that benefit could rise to between 5 and 5.5 per cent of gross national product. The scale of these optimistic forecasts suggests the great national benefits at stake if all should proceed according to plan. That the world energy market, the Western industrial system and, more immediately, the operations of the OPEC cartel may put all of these benefits at risk is a very serious caveat which must be registered, a caveat for which this book provides some historic as well as contemporary evidence. Some of these uncertainties arise from circumstances largely beyond the control, at least the direct control, of the British Government and people. Others arise from uncertainties of the British Government's own making. It is the latter and the nature of the political framework under which the energy industry operates which will most concern us not only in this introductory section but throughout this chapter.

(b) *The 1974 White Paper*

The present phase in the operations of the North Sea, or at least the outlines of the present political framework, derives from the July 1974 Labour Government White Paper which involved three principal proposals in the spheres respectively of government participation, corporation tax and regulation which we may briefly examine as to their terms in turn.

The chief means of government participation was to be through a British National Oil Corporation which would have the right to take up to a 51 per cent shareholding in the operating companies in the North Sea, and the corporation tax proposals (which effectively eliminated artificial losses which had been used to eliminate future tax liabilities) and proposals to separate the North Sea from other areas so that capital allowances on non-North Sea investments could not be used to reduce tax payable on North Sea activities. Finally, in the area of regulation measures, the most important provision of the White Paper was the government power to remit royalties. The power to receive royalty 'in kind' really related to BNOC's control over the crude oil. This was among the most significant items in the White Paper since the control over crude by an integrated company has much to do with the profitability of its refining and distribution system. It also represented an implicit assertion of national sovereignty over Britain's crude oil supplies in relation to the European Economic Community.

(c) *The British National Oil Corporation*

The provisions for creating BNOC as it has become known, were introduced in a Petroleum and Submarine Pipelines Act which granted BNOC broad powers to engage in the oil business including the power to operate pipelines, tankers and refineries. Many of its provisions applied to existing licences and were in consequence retrospective in character. To re-inforce BNOC's position the Department of Energy was also empowered to specify reasonable exploration programmes by private licensees with the implicit threat of BNOC participation if these were not fulfilled.

There are several major weaknesses already apparent in the structure of BNOC which need briefly enumerating. First, it has been given twin roles which are not logically compatible, (a) as an integrated national oil company and (b) as a general overseer of the oil and gas industry in the North Sea. Summarily, the argument runs that the knowledge that BNOC acquires in the second role gives it an impossibly privileged position in the first. Thus, under present legislation both BNOC and the British Gas Corporation are permitted to choose permit blocks without invitation, ie under the fifth round of licensing the three best blocks can be withheld for BNOC and BGC. Secondly, the voluntary agreement between British Petroleum and BNOC which guarantees no gains at the other's expense, thanks to the prior knowledge of BNOC, means that losses will tend to be centred on other companies. Furthermore, BNOC's agreement with various companies gives it increased knowledge of BP's competitors and thereby increased leverage on BP itself. All in all the creation of BNOC can be seen to be more about an assertion of tight political control over the oil companies, virtually an ideological statement, than extracting the economic rent which can quite effectively be done through royalties and corporation tax under present circumstances. This brings us to our fourth and final introductory factor which is a general principle rather than a specific measure.

(d) *Interventionism*

With the possibility of energy self-sufficiency by 1980, two broad policies present themselves to the government: either a policy of substantial energy exports or of a much slower rate of development, ie a so-called depletion policy. As this chapter later makes clear, the benefits of an export policy are likely to be considerable not only for Britain but also for her partners in the Community. However, a depletion policy of some sort is arguably desirable in Britain's and even the Community's longer-term interest, given the uncertain nature of the alternative fuels and the apparent inability of most Western industrial countries to slow the rate of energy consumption.

But while a depletion policy might be beneficial if it could be clearly formulated and applied, the energy industry is against any pattern which would develop productive capacity and then have it artificially restrained. At the moment the government has promised no restraints above 20 per cent of production, at least until 1982. This brings up the cardinal principle of the degree of intervention which is desirable for the effective exploitation and control of the North Sea.

At the moment the Department of Energy maintains that if there is a variation of more than 5 per cent in predictions of production (as submitted in the company work programmes) then the state has the automatic right to intervene, even to revoke licences at its discretion. Such a policy represents a very high degree of interventionism by any standards.

In response the companies claim that a depletion policy is in effect already in operation by default through the high degree of uncertainty, for which the swift and inevitable consequence has been delays, lack of incentive and caution by energy companies. The precise nature of these delays is enumerated later in this chapter.

In a word, BNOC seems to personify both in statute and

in practice all those elements of interventionism which have characterised the postwar history of Britain under both Conservative as well as Labour Governments, which we are now being obliged to re-examine. The solution may not be so drastic as a wholesale abolition of bodies like BNOC but more probably a willingness to fail to fully implement its truly enormous powers and to restrict its activities very largely to upstream operations.

For behind this very preliminary discussion of the framework for British energy lies the most fundamental of questions for which a solution must be found if Britain is to experience an industrial regeneration. Namely, how to create the framework for energy policy from a durable concensus among the people best qualified to know the requirements of the industry. As long as energy decisions are made giving only secondary consideration to commercial criteria, so long will Britain put at risk one of its greatest natural assets. In its suggestions for an Energy Advisory Commission together with a new Select Parliamentary Committee on Energy this chapter will attempt to respond practically to an enduring British dilemma.

Meanwhile the BNOC proceeded during its first year to buy its way steadily into the North Sea spending £396 million and raising a further loan of £407 million on the Eurodollar market. While its chairman, Lord Kearton, promised expected profits by 1979–80 it registered a loss of £1.2 million in its first year, ie 1976–7. BNOC's chief item of expenditure in its inaugural year was £287 million spent in acquiring the National Coal Board's North Sea assets in addition to a substantial share of Burmah Oil's offshore interests. The remaining £109 million went on exploration and contribution to the development of five oilfields and one gas field which the corporation acquired from the NCB and Burmah. During the same year BNOC and the Department of Energy also negotiated a 51 per cent state participation in most of the commercial oil fields in the North Sea. By 1980 these participation options plus the corporation's equity share in five oil fields — Thistle, Ninian, Dunlin, Statfjord, and Murchison — and a further option to buy any surplus oil from the British Gas Corporation, will give BNOC access to 30 million tons a year — equivalent to about a third of Britain's likely requirements by that date. Lord Kearton added that the corporation was devising plans for marketing substantial quantities of oil from 1978 onwards. In other words, the worst fears expressed earlier in this chapter were being fulfilled as BNOC began to make its influence pervasive throughout the North Sea energy industry. By its increasing monopoly knowledge of commercial information, BNOC was bound in the long run to inhibit competition.

5.2 A MAXIMUM TARGET RATE

In order to get quickly into the issues of central importance, it is useful to predicate a set of broad projections based on the most recent Brown Book produced by the Department of Energy, that entitled 'Development of the Oil and Gas resources of the United Kingdom, 1976'.

The fundamental assumption of this document is that oil and gas will remain Britain's principal energy sources throughout the 1980s; later in this chapter the debate hinges on who will be the principal suppliers in the 1990s and thereafter — imported oil, additional coal or nuclear capacity, or new sources such as solar, wind or wave technology energy. It is the existence of relatively plentiful oil and gas in the North Sea that has provided Britain with the excuse to delay making investment and general strategic decisions for the apparent energy gap which will arise when these deposits begin to run down sometime during the 1990s.

Basing our calculations on the estimates and evidence of the 1976 Brown Book, it can be said in very round figures, prior to the December 1976 meeting of the OPEC states, that British oil consumption was worth in the region of £4000 million, with natural gas valued at a further £1000 million. Thus, if the present rate of development is maintained making Britain self-sufficient in oil (she is already virtually self-sufficient in gas) by 1980, this would represent a saving on the balance of payments at current prices of £5000 million. In terms of income accruing to the British Treasury, basically royalties and taxes, at the current rate of around 85 per cent this would bring in around £4250 million on current projected rates of development. If the *right framework were provided*, during the 1980s the Brown Book estimates that the British North Sea and Continental Shelf generally is capable of providing exports in oil in the area of 50 million tonnes throughout the 1980s (at constant prices worth £2000 million) and gas in the vicinity of 80 billion cubic metres through the 1980s (at constant prices worth £1000 million), together representing an annual accretion to the Treasury of £2550 million. Thus, in broad summary, the Brown Book suggests that throughout the 1980s gross annual turnover from oil and gas is in the vicinity of £8000 million of which £6800 million would accumulate to the Treasury. As D. I. MacKay and G. A. MacKay's *The Political Economy of North Sea Oil* underlines, though this may represent an unprecedented transfusion of capital in the 1980s, it represents an extremely small employment creating effect. So much for those who still persist in proffering North Sea oil and gas as the latest in a long line of instant panaceas to Britain's poor postwar economic performance. But even these optimistic figures, which we have bracketed as a maximum target rate, are forecasts subject to major provisions of which the most basic is the provision of a stable and sustained yet still flexible framework which will encourage the required investment to be forthcoming.

5.3 A CRITIQUE OF THE CURRENT FRAMEWORK

To understand the balance of power in the political economy of the North Sea where Britain's currently exploitable principal energy resources lie, it is necessary to appreciate the scale of the revolution which took place in the relationship between oil companies and governments throughout the world during the 1970s. This revolution reached its most active phase to date in the period from 1972 to 1974 when by a rapid series of nationalisation measures and unilateral changes in contract terms, the Middle East OPEC states overturned the existing order leaving the oil companies faced with a *fait accompli* on prices, taxes and access to crude oil. The effects of this revolution, which is still manifestly going on, had a profound influence on the attitudes of European governments, and most especially Britain and Norway, in their handling of the framework under which North Sea oil and gas development has taken place. In particular the British Government perceived that the bargaining position of companies and governments had been irreversibly altered allowing the government to adopt a much tougher stance than it had hitherto contemplated. The consequences in the shape of tougher financial terms and unilateral and retrospective changes in licensing arrangements are not, in the circumstances, so very surprising. On the other hand, whether the tougher climate for oil companies has actually furthered the national interest is a subject worth scrutinising. Further strengthening the government's determination to extract a much harder bargain than hitherto, the quintupling of crude oil prices had made existing North Sea licences potentially much more valuable to the oil companies.

To begin at the beginning, the first round of government licences granted by the British Government to the oil companies was in 1964 when the then Conservative Government introduced the Continental Shelf Act under which certain specified areas were opened up to production. The criteria for awarding these licences however were not stated in the Act but were announced directly to the House of Commons by the Minister of Power. They included five principal conditions:

First, the need to encourage the most rapid and thorough exploration and economical exploitation of petroleum resources on the continental shelf. Second, the requirement that the applicant for a licence shall be incorporated in the United Kingdom and the profits of the operations shall be taxed here. Thirdly, in cases where the applicant is a foreign-owned concern how far British oil companies receive equitable treatment in that country. Fourthly, we shall look at the programme of work of the applicant and also at the ability and resources to implement it. Fifthly, we shall look at the contribution the applicant has already made and is making towards the development of resources of our continental shelf and the development of our fuel economy generally. (Hansard, House of Commons Debate 897 [1964])

As we shall see, each of these criteria was to have considerable bearing on the exploitation process of first gas and later oil in the North Sea for the next decade and beyond. At the very heart of these criteria was a pre-eminent political concern: on the one hand foreign companies with their capital, skills and manpower were needed to exploit the potential resources; on the other there was an underlying concern to favour domestic oil companies. The fifth condition in particular could be very easily interpreted to favour domestic applicants. In the first round of licensing 30 per cent of the blocks went to British oil interests and a further 10 per cent to Canadian interests. Whether, as has been suggested, more than 30 per cent of the most sought-after blocks went to British interests is unproven.

The means of payment for the privilege of exploiting what was still a largely unknown quantity was essentially a 12.5 per cent royalty on the value of production and more importantly the regular corporation tax. Altogether this amounted to a 50–50 division of the profits which was fairly standard worldwide at the time. As long as the prospective profits were uncertain it seemed a reasonable discretionary arrangement. The second round of licensing took place in 1965 and was essentially the same as the first round except that, this time conducted under a Labour Minister of Power, a further criterion was added: 'I shall also take into account any proposals which may be made for facilitating participation of public enterprise in the development and exploitation of the resources of the Continental Shelf.' (Hansard, House of Commons Debate 1579 [1965]) As a result, first the Gas Council and then the National Coal Board increased their participation under the second round until the Board's share of licences rose from 30 per cent in the first round to 37 per cent in the second.

The third round, which took place in 1969, saw further involvement by the Gas Council which formed a subsidiary, Hydrocarbons Great Britain Limited, to conduct its role as an operator, substantial gas finds having been made by this time. While each of these rounds witnessed the increasing participation of the state, they were each conducted under what can be loosely termed a discretionary system of licence allocation as distinct from an auction system. It is on the respective merits of these two systems that it is now necessary to turn our attention if we are to relate to the current debate about the relative efficiency of the current energy framework.

The British North Sea oil and gas fields have throughout, with very marginal exceptions we shall examine, operated under a discretionary system. In essence the system is based on granting licences in the first instance to those companies who promise the most active work programme or those who have previously demonstrated a very active work programme. In other words, the assumption has been that one of the prime requirements of the system has been to extract the oil or gas at the fastest possible rate. Less explicit is the assumption, though no less strongly held, that the discretionary system allows the government a much closer surveillance of operations, more particularly the power to discriminate in favour of domestic companies without it being subject to proof.

Advocates of this system argue that it logically must produce oil faster than an auction system since under the latter the bidder must put up money into advances to the government rather than into drilling. In return, advocates of the auction system maintain that by going to the highest bidder it goes to the best equipped and most efficient companies, ie those who can raise capital for bids do so because they can also raise capital for exploration and drilling. These arguments have an especially topical ring in the light of the claims in December 1976 of Professor Peter Odell and Dr Kenneth Rosing who alleged that oil companies' development schemes were depriving Britain of both revenue and jobs by their unnecessarily slow rate of development. Apart from the arguments for a depletion policy, which can be examined later in this chapter, the effects of over-competitive bidding on the basis of work programmes for blocks believed to be valuable may prove to be technically, and therefore financially, very wasteful.

5.4 AUCTION EXPERIMENT

The best means of making an assessment of the respective merits of the discretionary and auction systems is to examine the results of the auction experiment conducted in the North

Sea in 1971 by the British Government, upon which the
government seems to have made up its mind beforehand
and whose results have somewhat unaccountably never been
officially and publicly analysed. The background to the auction
'experiment' was the fourth round of licensing in which 267
blocks were allocated by the discretionary method and 15
blocks put up to auction. The only substantial qualification
to a 'pure bidding' auction was the government's stated
right to reject tenders who lacked the necessary technical
resources. This stipulation was included merely to exclude
purely financial interests interposing their services and thus
increasing the costs of development and presumably weakening
the degree of direct government control.

The stark financial figures demonstrate very clearly that
the auction was an unqualified financial success for the government. The fifteen blocks raised a total of £37 million, compared
with £30 million for the cost of the work programmes for
the 127 blocks allocated in the second round and £34 million
for the cost of the work programmes for 106 blocks allocated
in the third round. (Committee of Public Accounts [1973],
133) Apart from the advantages of immediate revenue to
the Treasury and the presumed auction winner's capacity
both technically and financially to exploit his licence successfully, these figures effectively undermine the chief argument
against the auction system. That argument has always rested
on the proposition that it is better for oil companies to
put their money into exploration than into advance payments.
But in the auction experiment the return on bids was on
average more than £2 million per block whereas the work
programmes cost only one sixth of that amount. While the
cost of the work programmes, due to cost escalation, rose
to very nearly £1 million per block, it represents less than
half the money obtained by the government immediately
from the oil companies in the auction experiment.

For those for whom this sample is incomplete evidence
the argument can be taken a stage further. Supposing that
companies would not embark on any exploration once they
had obtained a licence under an auction system — a truly
incredible assumption, yet sometimes argued by advocates
of the discretionary system — in which case the discretionary
system would be held responsible for all exploration included

in work programmes. It must be recognised that a discretionary system is an extremely expensive form of subsidy toward exploration costs. The key question is how much exploration activity could have been purchased for £37 million. If the answer is the amount expended on those fifteen blocks by a discretionary system, the figures are illuminating; using averages from the first three rounds, the amount would have been less than £12 million worth of exploration, and from averages derived from the fourth round, £15 million. Thus the discretionary system provides a subsidy twice as much as the companies will spend — in order to induce exploratory expenditure.

5.5 FAVOURING DOMESTIC INTERESTS

The fundamental reason why successive British Governments have opposed the introduction of any sort of auction system is not to accelerate exploration, though that may remain their declared aim and even intention at an earlier stage, but to enable it to favour British enterprises. Since the work programme 'bid' by any successful applicant is rarely revealed to the public, nobody can be sure to what extent it was the deciding factor in the awarding of a particular licence. The statistics which are available indicate their own story. Namely, that when the British Government had no opportunity to favour British interests, as in the auction experiment, the British share of awards fell.

In the Licensing Rounds Table provided in the First Report of the Committee of Public Accounts, North Sea Oil and Gas, Session 1972–3, the picture is quite clear. British interests gained only 22 per cent of the total area in the auction compared with 44 per cent in the discretionary part of the fourth round and between 30 and 37 per cent in previous rounds. The British ministry concerned has always been quite open, in fact, about the difficulty of favouring British interests being one of its objections to the auction system. If a British participation in any bid were to be made obligatory, it would be a clear violation of the Treaty of Rome, hence the need for a discretionary system to make such favouritism less explicit. That any government has the right to discriminate in favour of its own national enterprises, whether public or private, is not disputed, but that a certain economic price must also be paid for such selection processes should not be shirked.

5.6 GAS MONOPOLY CONSEQUENCES

One of the most fundamental questions in the examination of the current political framework in which the British energy industry operates is whether the use of the Gas Corporation's (formerly the Gas Council) monopoly has affected the willingness of companies to explore and develop new gas fields. In this matter exploration is for technical reasons especially important, since production from a gas field declines over time as pressure in the reservoir falls, making it imperative to discover new fields in order to maintain a steady rate of supply to the gas distribution point (Kenneth W. Dam, *Oil Resources*, p.73–86).

The answer to this question is by no means beyond dispute. At the same time the weight of what evidence is available, which is admittedly of a broad character, would indicate that the British Gas monopoly has generally speaking exerted a substantial disincentive influence. The evidence is broadly of two types: first, a decrease in the rate of exploratory activity and secondly, a willingness of the Corporation to increase very substantially the prices it pays for North Sea gas. Both of these factors need examining in greater detail.

First, the decrease in the rate of exploration. Ever since 1969 there has been a deceleration in the rate of exploration in the Southern Basin where companies looking for gas as their principal discovery would go hunting. The reason proffered for this decline in gas exploration is that companies quite simply found it more profitable to look for oil in the north rather than gas in the south. This might be a comforting explanation if it were not for the fact that international oil companies operating throughout much of the world are quite capable of optimising their rigs' locations, and if both northern and southern regions are assumed to be profitable to explore, the companies could be expected to explore in both simultaneously. Another argument, deployed by the

Department of Trade and Industry in 1972, was that the remaining unexplored structures in the Southern Basin were very unpromising. It is true that no major gas field has been discovered in the Southern Basin since 1968, but it is also true that the success ratio — that is, commercial finds to exploratory wells drilled — has noticeably improved. This may of course partly arise from greater familiarity with the geological structure of the area and similar knowledge accumulation factors.

But there is evidence that prices paid by the Gas Corporation or its predecessor have been the crucial factor slowing the development of gas. Thus Gulf Oil was unwilling to develop its Rough Field, which it discovered in 1968, at the then prevailing price and it was not until after a 1973 assignment to the Gas Council/Amoco group, which received a considerably improved price than had been offered to Gulf at an earlier date, that the field was opened up.

Further general evidence of the decreased rate of exploration in the Southern Basin, almost exclusively gas, compared with the Northern Basin, arose in the 1972 discretionary round of applications when only 34 per cent were taken up in the South compared with 77 per cent in the North.

Finally, amongst the objective criteria available — oilmen, it need hardly be added, are virtually unanimous that the gas price is generally speaking uneconomic at this stage, certainly to look for and develop new fields — is that the exploration of the Dutch sector of the North Sea has held up much better than the British. Although it is true that oil as well as gas has been found in the Dutch sector, the oil has not been in sufficient quantity to explain the relatively better rate of exploration. By contrast with the British, the Dutch gas price is relatively uncontrolled.

The second major reason why the evidence suggests the Gas Corporation monopoly has slowed development (long before talk of depreciation rates and other policies rationalising the events taking place were promulgated) is that 1968 prices slowed the rate of exploration as can be seen in the progressive price rises which were necessary before various fields were fully developed. Some of this price increase is due to the increased cost of development and inflation but this does not contradict a noticeable upswing in the price

paid per therm. Thus the price paid by Conoco/National
Coal Board for gas from the Viking field in 1971 was 1.5p
per therm; by 1974 the Gas Council contracted to buy gas
from the Rough Field at a reputed 3.4p per therm and
although its transmission costs were low (it required only
a 16 mile pipeline) the field was relatively small, even marginal.
A novel explanation of these particular price rises, above
and beyond the general increase in prices, was that the Gas
Council was attempting to pre-empt a Conservative Government's reported willingness to entertain direct contracts
between companies and industrial consumers.

5.7 INTRODUCTION: ENERGY POLICY PRINCIPLES

In the earlier part of this chapter we have proffered a very summary critique of how the British energy sector is working at present in the North Sea, illuminating in the process, we hope, the
fact that the present political framework for energy is deficient
in a wide variety of its aspects. In response to this general
picture it would be consoling to advocate a radical programme
of denationalisation in which the energy sector, heavily
dominated by nationalised industries and state shareholding
in the private sector, would be opened up to the winds
of competition. There cannot be much doubt that the progressive loss of competitiveness, the relative and steady decline
in Britain's economic position in which she has fallen from
third to twelfth position during the last twenty-five years,
is due in great measure to a loss of domestic competitiveness
to which the nationalised industries have substantially contributed by shouldering a wide range of non-commercial objectives.

However, not only has this system been steadily built up
over thirty years, but the even more time honoured special
role of trade unions in government (going back at least sixty
years to the Treasury Agreements of 1916) precludes any
immediate dramatic solutions of this sort. To take the most
obvious example, the possibility of denationalising the coal
industry, however much in reality it might secure the long-term
future of most miners, would be faced with a National Union

of Miners understandable intransigence wielding as it does monopoly bargaining power. As the National Coal Board is after central government and the post office the largest employer of labour in Britain, as its chief product is indispensable to the maintenance of the nation's industrial lifeblood and especially steel and electricity, and, not least, because it has dramatically demonstrated its political muscle, such options are clearly excluded.

There is nonetheless a clear necessity for Britain to evolve a framework for energy which will utilise more effectively the energy resources which are likely to become available during the next two decades. Such a framework will ideally be able to provide the public with a much less blurred set of choices of how Britain's energy resources can best be developed and reconciled with a host of political and social objectives. Hitherto, under a postwar momentum of generally subordinating commercial criteria whenever it conflicted with regional, social and increasingly sectional interests, the possibility of comparing the performance of a heavily nationalised energy sector with any alternative system was inconceivable.

The general economic declension, sometimes averred as the de-industrialisation of Britain, has however made another alternative just conceivable. Namely, an alternative framework which enables the presentation of the broadly commercial criteria for energy policy as a starting point to which specifically political and social amendments have to be argued in public on their merits. This is a sweeping aim and can only be demonstrated by the enunciation of such a programme of proposals step by step. The preliminary to taking such steps is to suggest a set of guiding principles.

The essence of an effective energy policy is to operate within an agreed framework of rules which are comprehensible and acceptable to all the principal parties in the energy industry, government, and not least, the consumer. It is with this in mind that this chapter recommends the adoption of six main energy policy principles as a basis for building a cohesive and attainable energy strategy for the future. These six principles are as follows: (1) Concertation (2) Accountability (3) Continuity (4) Flexibility (5) Security and (6) Collaboration.

5.8 CONCERTATION

The basic means to concertation recommended in this chapter is a reconstituted Energy Advisory Commission which would be a severely pruned down version of the one created in June 1977 (which comprises 22 members and is chaired by the Minister for Energy) more akin to Japan's Advisory Committee on Energy than its present British model. Unlike the present Energy Commission, the reconstructed one would be chaired by a wholly independent chairman and the Energy Minister of the day would assume a merely ex-officio observer role on the Commission, more often than not being represented by the Permanent Secretary from the Department of Energy. The new Energy Advisory Commission (EAC), would be a permanent standing body, financially and statutorily independent though working closely with the Department of Energy in drawing up a long-term energy strategy subject to continual revision. While it would contain a minority of representatives for consumer interests these would be provisional until such time as the energy sector as a whole had been opened up to free competition between its various components on the presumption that competition reveals consumer preferences more equitably than any other method so far devised.

The new Energy Advisory Commission would be composed of ten ordinary members plus a chairman, who would also serve as Director-General of the Energy Secretariat. The ten ordinary members would comprise the chairmen of the National Coal Board, the British Gas Corporation, the Atomic Energy Authority, the Electricity Council, the Petroleum Industry Advisory Committee, the Offshore Operators Association, the British National Oil Corporation (BNOC) and three representatives of consumer interests. There would be a conscious and notable absence of union representatives, not simply because the adoption of Bullock-style management with unions having an equal voice with management is likely to be a recipe for disaster, but because it has to be assumed that the chairman of the National Coal Board will take the fullest account of the interests of the National Union of Miners, and every other chairman of the union members in his respective energy sector. To allow unions to take up a third of the membership of the current Commission not only makes it unwieldy but so weighs it in favour of employment retention (regardless of its commercial viability) within the respective energy sectors that its credibility in advocating an energy policy in the national interest can be discounted. The reconstituted Energy Advisory Commission, by contrast, is roughly representative of the balance of interests within the energy industry, embracing five state corporation chairmen, two private sector chairmen and three consumer representatives, with an independent chairman of international stature to hold the ring between them.

The Commission would generally meet once a month to discuss and reconcile the conflicting components in the national energy sector. They would be served by a permanent secretariat of qualified specialists nominated by their respective Commissioners. They would, on advice from the Treasury and the Department of Energy, draw up their broad energy strategy within the parameters of an agreed budget, the priorities of which they would be under obligation to settle. Since through the Commissioners and the permanent delegates to the Commission they would be voicing the viewpoint of the major components in the energy industry, there is a good possibility that their deliberations would be well-informed. Should the proposals of the MacIntosh Report be introduced, then each of the Presidents of the Policy Councils on nationalised energy industries would qualify as Commissioners. Indeed, such a system would be complementary to the Policy Councils dealing with policy formulation, rather than management and the Commission providing the forum where the respective Policy Commissions could both inform themselves regularly about each other and increase the possibility of devising a long-term strategy for the entire energy sector. Fundamental to the whole concept of the Energy Advisory Commission is the necessity of creating the forum for drawing up long-term energy strategies, initially removed from the short-term pressures which are an inevitable concomitant of energy policy formulation deriving almost exclusively from the Department or the present Commission whose interests are so widely diffused as to give the Minister and his Department officials a dominating influence.

5.9 ACCOUNTABILITY

The first section of this chapter is a reminder of needless intervention into not only nationalised energy industries but intervention also into the general environment in which the private sector operates within the energy field. In order to act as a general watchdog of the public interest it might be valuable for a permanent Select Committee on Energy to be created with sufficient political and financial support to make it an effective investigative organ. The creation of a Select Committee on Energy would release the Select Committee on Nationalised Industries to engage in a more detailed scrutiny of the ever expanding public sector outside of the energy field. The Select Committee, which would have the power to hold hearings in public or private as it saw fit, would have the duty to examine in at least four or even five separate quarters, namely, the Minister for Energy and his departmental officials, each of the nationalised energy industries, the British National Oil Corporation and not least the Energy Advisory Commission. The Commission, once it had established its role and reputation, would hold out great possibilities for a touchstone on the national interest in the energy sector which would create the basis upon which the Department would have to justify its intervention and the various energy industries would need to account for any departure. The Commission itself, less it became too powerful, would be under continuous scrutiny by the increasingly expert Select Committee.

5.10 CONTINUITY

At present the permanent officials of the Department of Energy are left in more or less control of British energy policy. That it remains still a very much ad hoc arrangement may not be the fault of the present officials given that energy policy (by its nature a long-term process running into five, ten and twenty year time scales) is irreconcilable with the change of minister effectively every two to three years. Governments themselves have on postwar performance a life expectancy of around four years so even if the Minister for Energy remains at his post throughout the life of his administration there is always the possibility that long-term planning will be set into reverse by the next administration. The continuity which is such an essential ingredient in effective energy strategy cannot realistically be provided from the Department of Energy unless conditions hitherto absent were to be introduced. The most important of these is a strategy which has arisen fundamentally from a debate, and a continuing debate at that, between the principal parties in the energy industry. The creation of an effective Energy Advisory Commission provides such a possibility and one which would provide the tool allowing the Department to argue for the maintenance of continuity unless the political arguments were overwhelming. Moreover, with a watchful Select Committee, it would be possible to examine whether these 'political' arguments were in fact consonant with the preponderant national interest and whether the sectional pressure to which the government might find itself exposed could be satisfied in another manner.

5.11 FLEXIBILITY

In outlining suggestions for concertation and the maintenance of continuity, it would be a fatal error to overlook the need to provide maximum flexibility within any new framework. The necessity for flexibility needs no underlining even under present circumstances where the variations in scarcity and price on the world oil market, and therefore indirectly the world energy market as a whole, have fluctuated alarmingly. As long as the OPEC cartel is able to maintain its united front at the crucial moments of decision, there will forever remain this fundamental unpredictability of both supply and price for oil imports, and therefore, since it still represents such a substantial proportion of the whole, for energy supply and demand generally.

The chief means to flexibility in the future is unlikely to be institutional. Indeed, it is the creation of a pattern of cross-fertilisation between the principal institutions concerned with energy production and administration that will become increasingly important. As the Macintosh Report revealed, the nationalised industries' board members and management commonly complained that civil servants in the sponsoring departments, and related departments such as the Treasury, did not generally seek the corporation's advice on policy issues and rarely consulted them while policy was being developed. It was suggested that civil servants were insufficiently aware of the consequences for an industry of changes in policy. The best antidote to this widespread compartmentalisation is much greater interchange of personnel between both corporations and departments.

5.12 SECURITY

So far in our proposals we have stressed various means to greater overall coherence in the national energy sector and of increased efficiency in its respective components. However, as the continued existence of the OPEC cartel, the widespread existence of both high inflation and unemployment as well as low growth (not least in Britain), the intractability of the problems between countries in the Northern and Southern hemispheres of whom the latter are largely economically the least developed, the intransigence of the parties to the Israeli-Arab dispute in the Middle East, to say nothing of their great power sponsors in that area, there is clearly no lack of raw material for conflict in the economic sphere alone even assuming no wars break out into the open.

The creation of self-sufficiency and security of supply is therefore a paramount consideration for a British energy framework as for every other industrialised nation. In purely British terms this is not too remote a possibility to contemplate with the likelihood of Britain attaining energy self-sufficiency by around 1980. The degree of economic and industrial interdependence which exists between all industrial nations and most particularly those of Western Europe has already been outlined in earlier chapters. How Britain is specifically prepared to respond toward her partners in the face of a possible renewed oil embargo sustained over a much longer period than hitherto is something which needs much greater clarification. The degree of co-ordination between West European states on energy matters is still woefully lacking.

5.13 COLLABORATION

Proceeding from the last principle of the paramount importance of security in an uncertain world, it is a natural step to move on to the advocacy of greater international collaboration. In Britain's case this falls into three more or less distinct areas. First, there is the sphere of international collaboration in the sharing and pooling of research. Here the International Energy Agency, of which Britain is a prominent member, is possibly the best placed vehicle to translate the verbal commitment of the US Government, for instance, to share its vast energy research resources, though nobody should linger under the impression that energy research is a one-way sharing process. Second, like both the United States and the Soviet Union, Britain is soon likely to be among those countries who are both major producers as well as consumers of energy. Britain, however, differs from the two superpowers and most other major energy consumer nations in that she is likely to be capable of developing a not insubstantial energy export potential. Much of this will depend on the effectiveness of her energy framework, not least her ability to restrain the growth of domestic energy consumption and develop more effective conservation measures both technologically and socially. It is as an energy exporting nation which is at the same time a major consumer that Britain may have a unique function to fulfil in the future in the dialogue between the oil exporting and oil consuming nations. Thirdly, and finally, the complementary nature of Britain's energy resources and the requirements of West Germany should be seriously pursued as the most realistic foundation and starting point in any attempts to create a greater degree of energy collaboration among the EEC member states.

5.14 INTRODUCTION: PRIORITIES FOR POST-1985

The intention of this third and final major section is to examine the possibilities for the most rational use of resources in the fields of coal production, nuclear power, conservation and renewable resources. Before analysing these energy sectors in any detail it would be useful to record the preliminary

strategy of the Department of Energy as presented in the
Energy Policy Review of January 1977.

5.15 CURRENT POST-1985 STRATEGY

The following summary of the Department's strategy for the
period leading up to the year 2000 must necessarily be
couched in the most general terms, not least because both
the balance of supply and demand towards the end of the
century and the contribution from each indigenous source
(as well as and even more from imports) cannot be predicted
at this point with any final degree of accuracy. Nevertheless,
decisions on the allocation of resources affecting this period
very substantially have to be made almost certainly before
the end of the 1970s at the latest. The following six points
(this author's numbering) seem to arise from the Department's
long-term strategy:

(1) to press ahead with the economic development of the
    coal industry and with coal utilisation and coal conver-
    sion research;
(2) to make Britain's approach to oil depletion policy as
    flexible as possible;
(3) to keep the long-term strategy for gas development
    under review;
(4) to encourage all cost effective steps to conserve energy;
(5) to ensure the availability of nuclear technology and
    the manufacturing capacity needed for re-equipment
    and expansion in the 1990s;
(6) to establish the viability of renewable resources.

As the Energy Review Paper notes, the total costs of such
a strategy will be vast and this is where a proportion of
North Sea oil and gas revenues should be set aside to meet
such costs.
    The Energy Review Paper makes a number of important
observations both in the field of world energy and British
energy that are worth recording, not because like the previous
six points they are anything but unexceptionable, but because
they help provide the background factors in which the options

available can be discussed. While at present there may exist
an excess of potential oil supply over demand, oil prices
are likely in general to remain high. This central assumption
is a principal reason why the nuclear power programmes
have been given such an enlarged function by so many Western
energy planners. At the same time, quite apart from safety
and environmental objections, the scarcity of uranium is likely
to become more acute and with it an accompanying rise
in price of that commodity. Since oil and gas production
in the North Sea is likely to be in decline by the last
decade of this century the crucial question for British energy
has for some time revolved around the balance between coal,
nuclear power and alternative energy sources, but especially
the former two with which we must now treat in greater
detail.

5.16 NATIONAL COAL BOARD PLAN TO THE YEAR 2000

Since coal is likely to be one of the major options in any
energy strategy for Britain post-1985 and since it requires,
like nuclear energy and most alternative energy sources at
this stage, long lead times for its effective development, it
is not surprising that the National Coal Board has already
mapped out a general proposed strategy to meet the energy
requirements of Britain's domestic and industrial markets up
to the year 2000. Setting the scene for its examination of
the possibilities for coal, the NCB Plan 2000 notes that world
energy demand has quadrupled since World War Two, that
it is likely to continue to do so especially in the light of
world population growth (this argument should not be over-
stated since most population growth takes place in low energy
consumption areas), and that with a growth rate of little
more than half that experienced in the last thirty years,
energy demand should still double. Moreover, according to
the Plan, if oil and natural gas are the world's scarcest
and most valuable energy sources and that they are being
the most rapidly depleted, the likelihood must be that the
'real' price of energy will rise substantially before the year
2000.

The Plan goes on to point out that while energy self-sufficiency will almost certainly have been achieved by 1980 in Britain this should not lull the nation into a false sense of security. It suggests that if a large coal industry were to be sustained and expanded it would enable the nation to (a) export oil, and possibly gas and coal from an energy surplus and (b) to conserve North Sea oil and natural gas by controlled depletion. Both (a) and (b) would be possibilities during the 1980s. Then, by the 1990s, if not before, there is a third argument, (c) to preserve energy independence, even taking account of nuclear energy and the impact of various conservation measures.

If such an expanded coal industry is to be achieved, the Plan argues, action will be required very soon since the time required for exploration, investment and research is considerable. After allowing for 'energy conservation', the requirements for primary fuels are estimated to increase as follows (related to a yearly gross domestic product growth of 2 per cent or 3 per cent): 1975 actual demand was 344 million tons of coal equivalent, 2000 low growth might be 490 *mtce*, 2000 high growth might be 630 *mtce*.

The Plan 2000 goes on to discuss the nature of the potential markets for coal. At present it notes the largest market for coal is power stations, for the future it will depend both on the sales of electricity and upon coal's share of the total power station consumption. The estimated power station fuel requirement by the year 2000 is 50 per cent above the present level with the fossil fuel requirement remaining broadly at the present level. Thus, the Plan argues, provided coal can be produced at acceptable cost, there are strong economic and energy policy grounds for reducing the amount of fuel oil and natural gas used in power stations and concentrating the fossil fuel requirement on coal, thus reserving natural gas for premium uses. As part of these arguments for coal-fired power stations there has been considerable recent controversy about the ageing capacity of the coal-fired stations, for no new ones have been built since Drax I in 1966. Unless there is a reversal in policy on power station construction, while generating capacity is likely to increase by 1985, coal-fired capacity is likely to fall. The Plan 2000, consistent with its overall view on expanding coal investment and capacity, argues for the replacement of coal-fired plant with new power stations, the modernising of existing stations and the conversion of oil and gas-fired stations to coal.

The Plan points out that in 1975 industry consumed about 67 million tons of coal equivalent of which, apart from electricity, coal provided only 11 million tons compared with 29 *mtce* for oil, and 18 *mtce* for gas. Thus coal provided less than a fifth of the direct use of fossil fuels in the industrial market. While the next ten years could be tough competitively for coal, by the 1990s, the Plan argues, coal should become the main fuel for industrial steam-raising, especially for large industrial consumers where coal can be handled in bulk. While a market for industrial coal of the range of 30 to 50 million is feasible according to the NCB Plan, such a reversal of the trend in industrial coal sales would be consequent on a significant price advantage over oil and a general improvement in coal equipment in industry. While the prospects for improvement in the latter case are reasonably promising thanks to the progress in coal technology not only in Britain but internationally, the possibility of coal maintaining a price advantage over oil is suspect.

5.17 COAL DEMAND PROSPECTS

While Plan 2000 sees great future prospects for coal conversion — whether fluidised bed combustion, coal liquefaction or pyrolysis — it also acknowledges that these will not become contributing factors in energy supply on any significant scale probably until the very late 1980s, by which time incidentally some of the alternative energies such as solar heat, wave and wind power will be making an expanded contribution. The Plan envisages the main markets for coal for the next two decades as consisting of: (i) bulk combustion at power stations and (ii) industry; also (iii) the use of coking coal for the steel industry. The general assessment of Plan 2000 is that the long-term markets for coal are likely to be as great as the present with the possibility of their becoming greatly expanded. The Plan provides the following predictions of coal's long-term production potential for the year 2000 with an upper and lower range of coal demand:

Power stations	75 million tons	up to	95 million tons
Coke ovens	20 million tons	up to	25 million tons
Industry	30 million tons	up to	50 million tons
Domestic/commercial (including SNG)	10 million tons	up to	30 million tons

The Plan includes an argument for regarding imported coal as no more than a subsidiary future source of coal supply on the grounds that internationally traded coal will show little advantage over expensive oil imports, assumptions requiring scrutiny later in this chapter. The Plan rightly points out the likely increase of British coal exports to the EEC from 40 million tons now to possibly 60 million by 1985, and rising thereafter.

5.18 COAL RESERVES, CAPACITY AND INVESTMENT

Britain's 'technically recoverable' reserves of coal are estimated by the NCB Coal 2000 to amount to some 45 billion tons which is enough to last 300 years at current rates of production and significantly several times greater than the ultimate combined reserves of North Sea oil and natural gas. Several points can be made about the nature of the reserves which, unlike the economic aspects of the coal extraction process, can be stated without much qualification. They are: (a) in terms of seam thickness, the classified reserves at existing collieries will enable the average working section to be broadly maintained, at least to the year 2000. (b) there are very large potential reserves generally in localities of low geological disturbance — mainly in extensions of the Yorkshire and Midlands coalfields. (c) the working depth of a major part of potential reserves is no more than commonly experienced in the most productive of existing collieries.

The mining environment of potential new mines is likely to be significantly better than that found in much of the existing capacity. Moreover, if the necessary authorisation is granted, the reserves of opencast coal are being proved continually and will maintain a yearly rate of at least 15 million tons well into the next century.

To make any assessment of the amount of investment required

to fulfil the NCB's Coal 2000 intentions, it is necessary to examine the requirements for both new capacity to meet the optimum (or near optimum) demand and, in such a long-term assessment, the rate of exploration. While any calculation about the exhaustion rate must be imprecise if it extends as far ahead as the year 2000, analysis of the reserves remaining at existing collieries indicates that the average yearly loss of capacity between now and 2000 might be 2 million tons, within a possible range of 1 to 3 million tons. Such an assessment assumes the continuation of the existing mining technology and the maintenance of coal's competitiveness in relation to oil. The Plan 2000 suggests that some 60 million tons of new capacity would be required by 2000, over and above the Plan for Coal programme, representing the commissioning of about 4 million tons a year from the mid-1980s onwards. The greater part of this new capacity would have to come from new mines with individual outputs of 2 million tons or more. The most likely estimate among those available would require the setting aside for potential development of about thirty new mine sites with at least twenty-five years life. This might involve an optimum potential of 1.5 billion tons of coal.

From the previous paragraphs it will be apparent that the expansion of British deepmined coal on a substantial scale in the period beyond 1985 is in terms of the reserves and the maintenance of the present rate of exploration perfectly feasible. The chief constraints are likely to be the availability of both capital and expertise. Thus to work toward a deepmined capacity of 150 million tons in the year 2000 would require first, continuing capital of £400 million annually to cover new capacity, replacement of plant and machinery and for technical advance. This is roughly in line with the current average level of yearly expenditure during the period of implementation of Plan for Coal which expires in 1985. Second, it would require sufficient assurances for the long-term framework and future of the industry for it to attract and retain the necessary human skills and create the climate for technical progress. It is the enormity of these financial demands for capital expenditure and the difficulty of being able to provide the assurances that the industry not unnaturally demands that constitute the nub of the question when it comes to approve or disapprove the general lines of the NCB's Coal 2000.

In a moment we will suggest the touchstone by which proposals of such far-reaching importance should be judged, but in the meantime some brief conclusions on the Coal 2000 prognosis are in order. The underlying emphasis of Coal 2000 is the central importance of guaranteeing British energy long-term independence against a background of global energy uncertainty, more especially the uncertainty of supplies of oil and natural gas. While Coal 2000 admits that market conditions for coal will be difficult until the mid-1980s because of the expansion of North Sea oil and gas, it argues that thereafter there will be scope for a significant expansion for coal. It is the general principles and presuppositions that undergird the arguments advanced in Coal 2000 that need now to be more closely examined in order to determine the wisdom of adopting, modifying or rejecting them.

5.19 PAST COAL POLICY

A great deal of postwar British economic policy has recently been, as it were, found out. The popular myth, for example, that it is possible to preserve the characteristics of a broadly free society and at the same time hold the government of the day responsible for the *provision* rather than the mere securing of the conditions for full employment has been rumbled recently. To this popular myth the British energy industry has more than paid its dues. A very important part of contemporary energy mythology, widely accepted by energy experts and sophisticated laymen alike, is that the energy policies of successive British governments over the last fifteen or more years prior to 1973 in running down the coal industry was shortsighted. The argument rests on the proposition that if coal output had been stabilised OPEC's current power would be less damaging to British energy interests and that it would provide the foundations for a long-term future when we could afford to ignore foreign energy sources.

The most obvious fallacy in this view is that the British Government did not consciously seek to run down the British coal industry; on the contrary they constructed a series of protectionist measures designed to slow the rate of decline in coal output and in manpower which is quite a different thing. As Professor Colin Robinson points out in his excellent paper The Energy 'Crisis' and British Coal, Hobart Paper 59, these measures were very comprehensive by any standards. They included (a) a tax on fuel and heating oils first imposed in 1961 (originally equivalent to 30 per cent of the industrial fuel oil price); (b) substantial preference to coal in electricity generation and to a lesser extent in the rest of the public sector; (c) the exclusion of overseas coal and Soviet oil at times when they were cheap enough to be marketed in Britain; (d) substantial direct government aid to the NCB in 1965 and 1973 when the value of its assets were written down; and not least (e) the 1973 Coal Industry Act which provided grants of up to £720 million over the subsequent five years. Taken together the evidence is overwhelming and conclusive that British energy policy has been consistently aimed at supporting coal in whatever way it could.

The logical consequence of maintaining that coal output should have been retained at its highest postwar level or thereabouts is an argument that the coal industry should have even more protection than in fact it enjoyed in the past. Specifically, from the late-1950s the government was being lobbied by the coal industry to stabilise coal output at around 200 million tons of coal annually. It is a fact that successive governments, to their credit, were unwilling to take measures that must have almost inevitably included stricter controls to make the Central Electricity Generating Board burn more coal, more subsidisation of coal, increases in the fuel oil tax and ultimately fuel rationing. Viewed from the wider perspective of the national economy it made no sense to have kept major resources of both labour and capital in the coal industry when they were needed in expanding sectors of industry. For, in spite of the protection already described, there was a massive switch away from coal to oil and natural gas during the decade and a half preceding 1973.

Moreover, if the target of maintaining a coal output of 200 million tons per annum since 1960 had been attempted, in 1973 a minimum of 140,000 more miners would needed to have been recruited, that is an increase of 50 per cent on the actual labour force. This figure, based on achieved output per man, is almost certainly a considerable underestimate since such a policy would have involved keeping open pits in low productivity areas. Apart from depriving other industries who might otherwise have attracted the same labour into unprotected employment, it is obvious that an increase of the coal mining labour force on this sort of scale would involve paying very much higher wages than those which prevailed.

What becomes increasingly apparent from the evidence is that a higher coal output, in effect a more protected coal output, had it been engineered by government policy, could only have been achieved by a major switch of resources into coal. For if Britain had denied herself the opportunity to import oil for most of the 1960s, substituting indigenous coal in its stead, this would have meant higher costs to manufacturing industry and eventually higher costs throughout the entire economy. The fundamental question that must be put is what, given the likely costs already described, would have been the likely benefits arising from such a policy of maintaining British coal output at a specified level by means of protection? It should be noted that the benefits which were claimed for a policy of protecting coal then were not too different from those advocated (with differing emphasis) for the future in Coal 2000.

The chief benefit accruing from a more protected indigenous coal industry has almost always rested on the question of security of supply, that is, of Britain's total energy supply. Thus by 1973, if 200 million tons of coal had been produced and consumed instead of the actual total of 130 million tons by means of protection, the proportion of indigenous fuels in total British energy use would have risen from 38 to 58 per cent, assuming the same total energy consumption. This, its advocate argue, would have constituted an unmitigated gain by securing a greatly improved security of supply by decreasing the amount of imported and presumed unreliable energy supply.

In a strictly strategic sense this may have been true, namely, that security of supply in wartime is normally ensured by reinforcing domestic sources. But in periods of non-belligerence there is nothing inherently more secure in home-produced

energy. On the contrary, if we compare the respective perform-
ances of coal and oil over a given period we discover
that in the case of oil there was no serious interruption
of supply between the time of the 1956 Suez crisis and
October 1973, while the coal disputes of 1971–2 and 1973–4
were quite as damaging as the Arab oil restrictions on total
British energy supplies. Moreover, and highly germane to
the proposals for massive long-term future investment, if coal
output had been higher than it was in the first half of
the seventies the effect would have been correspondingly more
severe. In retrospect therefore, coal has not been more secure
than oil to date; the continuing presence of monopoly bargaining power in the hands of the miners, reinforced by the
inability of government to reduce it, virtually guarantees
built-in insecurity of coal supply unless the government capitu-
lates to permanently extended protection for coal.

The subsidiary argument for coal protection revolves around
the point that, despite increased production costs for domestic
energy in the period from 1960 to 1973, by protecting an
indigenous fuel such as coal Britain might have made temporary gains from lower industrial fuel costs. While it is true
that with greater protection, for a time coal would have
(at least artificially) regained its previous competitiveness over
the increased price of oil, and it is also true that post-1973,
as oil prices rose further, this would boost coal's competitiveness, it must be a reasonably safe prediction that massive
wage increases would have eventually brought coal back to
parity with oil. But even this apparent short-term gain
is probably illusory. For if British coal had been more available,
the OPEC powers would almost certainly have imposed even
tougher restrictions, that is, they would have enforced stricter
quantity controls on supplies of higher-priced oil. Any other
conclusion implies a state of competition which is notably
absent in the spectacular rise of OPEC.

Furthermore, it is not only the price of oil which would
have been affected by substantial British coal protection prior
to 1973. With increased protection it is reasonable to assume,
on all past evidence, that coal prices would have been higher
than they proved to be, lacking the pressure toward more
efficiency provided by low oil prices in the 1960s. The
labour and capital costs of expanding production under such

a regime would have been very large indeed. Last, but not least, the miners under such a system would have been in an even more unchallengeable position, leading more or less inevitably to higher wages as soon as oil prices rose. This indeed happened in fact though it would have been more irresistible and costlier under a more protected system.

5.20 FUTURE COAL POLICY

On the available evidence examined in the previous section, Britain would have in the long term benefited from less rather than more coal protection. In the following section we will examine how future coal policy might be conducted and especially the guiding principles for an effective coal policy for the long term. As a preliminary step it is necessary to put coal into the context of its relationship with other fuels.

While in the past energy policy was heavily preoccupied with the competition between fossil fuels, coal and oil in the 1960s, and additionally, natural gas in the seventies; in the future it is likely to be different in a number of important respects. First, oil will become an indigenous as distinct from imported fuel; second, natural gas will become more plentiful for a period of some years at least; and thirdly, nuclear fission will become for the first time a major source of energy for electricity generation. In the latter instance, though major investment decisions affecting the 1980s and nineties need to be made now (as we shall see later in this chapter), nuclear power will represent a still quite minor source of total British energy supplies for the remainder of the seventies. Thus the fuels most affecting coal will remain oil and natural gas, more especially in the period of the 1980s when the indigenous supplies of both oil and gas are likely to be relatively plentiful.

While there is still room for debate about this assertion, there is evidence to suggest that total energy consumption in Britain will probably rise relatively slowly during the remainder of the seventies; this has a great deal to do with the fact that an increase in energy prices relative to prices in general will provide a real incentive to economise quite apart from any specific effects of conservation measures. Taken together, the slow rise in total energy consumption alongside a substantial increase in natural gas and nuclear electricity could represent a reduction in the combined consumption of coal and oil. If we take the base year as 1974, when total energy consumption was around 340 million tons of coal equivalent, and energy demand increasing at 1 per cent per annum up until 1980, it would rise to approximately 370 million tons of coal equivalent by 1980. A major factor in whether the consumption of both coal and oil may significantly decrease as a proportion of the total energy consumed depends on the pricing policy for natural gas.

Around the world, natural gas has proved itself one of the most desirable fuels both in terms of its convenience to handle and transport and its technical versatility. When priced at thermal parity with other fuels, natural gas has rapidly superseded its rivals in many markets. This very competitiveness has led to some spectacular policy failures of which the US Federal Power Commission's holding down of the price of US gas (which led to over-stimulation of demand by artificially low prices) is the most notorious example. The consequences in the shape of severe shortages and lack of incentive to explore and exploit became most pronounced in the winter of 1976–7 when both industries and schools were forced to close because of the shortage. For a very long time, as we saw in the earlier part of this chapter, British gas from the North Sea seemed to be charting a similar course of over-stimulation of demand by government price policies as exerted through a state monopoly which, as constituted, had very few criteria upon which to calculate the market value of gas.

The background of thirty years of a state-dominated energy sector has created a climate in Britain that regards it as legitimate for the various energy sectors to pass on to consumers, sometimes, cost increases but immoral to increase prices to take advantage of demand conditions. This belief is at the root of much of the overall inefficiency and relative poor value for money which the respective state energy sectors have provided despite the array of qualified and competent people to run these industries. Thus while three of the fuel industries in the shape of oil, coal and electricity could claim their costs had increased and thus put in for price increases, the gas industry was instructed to hold its prices down since it was neither at the mercy of oil producers, strong unions nor, in the case of electricity, the two in combination. While most recently there have been signs of the slight raising of gas prices so that they are not so strikingly out of line with coal and oil, and despite the current and future flow of 'associated gas' (ie linked with oil) from the northern offshore oilfields, it will be of paramount importance not to set the price so low that excess demand will be created and a great national asset prematurely squandered.

Putting gas policy aside for the moment, we come face to face with the crucial factors which most intimately affect our attitude toward future coal policy, namely its relationship with oil. As all the world now knows, Britain is fast moving toward self-sufficiency in oil, with a potential output by the early 1980s of the order of 150 million tons which is somewhat in excess of Kuwait's production in 1973. By 1980 imports from OPEC will be balanced by exports of North Sea oil ushering in a period of potential oil exports on a substantial scale. The important point to register is that oil will have become as much an indigenous fuel as coal, thus putting a new complexion on the arguments for protecting coal against oil to create energy self-sufficiency.

Even if the United States were to establish energy self-sufficiency in the foreseeable future, which is somewhat improbable as of now, it would be at the cost of raising the overall costs of US energy by at least a third its present level. By contrast, if Britain were to introduce by stages a measure of free competition between her three indigenous fossil fuels — oil, gas and coal — she could both become and remain self-sufficient in energy until her national renewable sources of energy had come into large-scale production. That at least is the ultimate possibility that beckons across the barriers of existing but obsolete barriers to free competition under which successive postwar British energy policies have in practice been conducted. To be able to move in the direction of free competition it will be an essential prerequisite to dispose of some of the arguments for protecting coal in the new context already described.

There were, as Professor Colin Robinson notes in his Hobart

Paper, *The Energy 'Crisis' and British Coal*, four types of benefits that have been suggested in the past as *externalities* or social benefits arising from coal which cannot be measured in its price.

(1) *Improving the balance of payments*

While producing extra domestic coal rather than importing coal might have made the balance of payments statistics better in the past they may well have affected energy costs adversely; it is unlikely that the overall balance of payments ever gained from this kind of addiction to trade statistics extracted from their context. Under the new situation where each of the main fossil fuels are indigenous, even these statistical illusions on the balance of payments are likely to disappear abruptly.

(2) *Security of supply*

As we have already argued, coal has been subject, and unless drastically restructured will again be subject, to more serious interruptions than oil over the last few years. Unless militancy develops among the workers on British offshore oil and gas fields they should by comparison with coal be much more secure sources of supply than coal, at least in the medium-term future.

(3) *Rising oil prices*

As far as OPEC prices are concerned Britain is not likely to have much political muscle; even the EEC is in no position to exert much leverage on the OPEC nations. To the extent that as a net exporter of oil in the 1980s Britain has an interest in keeping the oil price at a reasonable level, she has, of course, a conflict of interest here looming on the horizon. The danger would arise if, seeking to increase coal output by a government-supported crash programme, it might be a five or even ten year crash programme which in a field of very capital intensive structures and long lead times could be unjustifiably, even cripplingly, expensive if it produced coal which ended up by being infinitely more expensive than the imported oil. For such are the time lags of the coal industry that there must be a possibility of oil prices, if not falling, failing to increase at the same rate as the capital costs of long-term coal investment programmes, not to mention the recurring labour costs. While it is not likely to drop in the foreseeable future it is as well to remember that the oil price presently paid to the OPEC producers bears very little relationship to the costs of production, merely the strength of the cartel. In other words, when the necessity arises, many of the OPEC countries are capable of accepting an appreciably lower price for their oil when the situation might demand it, as when alternative energy like domestically produced coal was proving too competitive.

Quite the strongest argument for coal protection, or put more positively long-range, large-scale investment in coal production, is that by the 1990s and beyond Britain may need a larger coal industry than the present market will produce. If you like, for Britain, coal is an insurance against something going wrong with nuclear power expansion when the oil and gas have passed their peak of productivity. This is, of course, a major ingredient in the argument for coal worldwide and in a global context it carries considerable weight. However, there is a need for a clear distinction between Britain's labour-intensive coal industry and that of the United States, for instance, where the opportunities for opencast and therefore cheap coal are still widespread, subject to no extraordinary increase in environmental legislation.

(4) *Social effects*

In the past the social consequences of an uncontrolled decline in coal employment which might have followed in the wake of a fully competitive energy market were among the chief foundations of state support for coal mining. But that is not the present option before the energy planners. If output is now stabilised employment will fall only gradually in response to a rise in productivity from new mines, better mining equipment and technology etc. The age of the majority of miners is quite advanced allowing a policy of natural wastage to be implemented fairly readily but posing a major challenge if a plan for expansion were embarked upon with consequent demand for a young, highly paid, skilled labour force to be recruited.

The extent to which the present coal industry is still subsidising uneconomic pits for the sake of employment can be seen quite readily by noting that of thirteen mining areas in the accounting year 1975–6 five of them made substantial losses, namely the Western (Cornwall-Devon) coalfields £12.1 million, South Wales coalfields £9.6 million, Scottish coalfields £8.9 million, Barnsley coalfields £3.5 million and finally the lowest loss was made by the Kent coalfields with £2.2 million; all other coalfields made a profit. This single year confirms the continuing picture of subsidised uneconomic coalfields in the Celtic extremities of Southern Scotland, South Wales and the Western (Cornish) peninsula.

Having examined the claims of some of the benefits arising from a policy of favouring coal rather than its competitors (whoever they might be at the time) it is worth briefly noting that there are additional *costs* to coal mining, too, in the shape of the environmental effects of both coal production and consumption, ie on the landscape and in atmospheric pollution, etc. One could argue that on both these grounds coal might merit discriminatory charges against it and in favour of either natural gas or even oil. Furthermore, though to run down employment in the coal industry too rapidly would be unjustified, there is a sense that since mining remains both a dangerous and unhealthy if not unpleasant job, at least at the coal face underground, it is desirable both for the sake of society as a whole and the individuals directly employed that other forms of energy should be single-mindedly developed and exploited.

But in summarising the arguments of principle over whether it is sensible and justifiable to extend state financial support to a greatly enlarged coal industry, the case is by no means proven. The most persuasive argument for pursuing a policy of coal expansion under the general lines advocated in Coal 2000 is the retention of capacity for the period beginning in the late-1980s when nuclear power will be beginning to displace oil and gas but when the demand for coal may remain quite substantial. There is nonetheless a strong argument for allowing the market as much scope as possible in determining the extent of the expansion in coal in a 'natural' manner, ie gradually, rather than pretending that

we have sufficient foreknowledge of the state of the demand
for the respective energy sources so precisely.

5.21 NUCLEAR POWER POLICY

Throughout much of this chapter the intention has been
to argue the case for competition as the best means to achieve
a rational and just energy policy upon which Britain's new
industrial infrastructure must be built. Because competition
implies mostly competition between Britain's three major fossil
fuels which will each very soon constitute indigenous assets,
by far the greater part of this chapter has been devoted
to coal and North Sea oil and gas. It would nevertheless
be useful to take some account of the other sources of energy
which will increasingly compete for investment and human
resources as time goes on and as new technologies are refined
to the point when they can make substantial contributions
to the total national energy supply.

By the year 2000 primary energy demand in Britain is
expected to reach 525 million tons of coal equivalent compared
with around 325 million tons of coal equivalent in the
mid-1970s. By arithmetical projection only it is expected that
there will be a supply of 175 *mtce* in oil from the North
Sea, 150 million tons of coal, 60 *mtce* of gas, 50 *mtce* of
nuclear power, and 35 *mtce* from various alternative sources,
giving a total energy supply of 470 *mtce*. This shortfall is
not too worrying if one takes into account some sort of
success for energy conservation legislation. But the significant
figure amongst the expected principal suppliers for the future,
in investment terms, the very near future, is that of nuclear
energy.

If we accept this very modest growth in potential energy
demand in Britain, among the lowest forecasts within the
range of probability, it would nevertheless require an almost
fivefold increase in nuclear power. For the alternative energies
the figure is even higher but since each of them — that is
wind, wave, tidal, geothermal and solar power — are starting
from an almost zero base rate, the figure is less daunting
though still staggering. The two figures placed in juxtaposition
are significant since they underline the lack of realism of

those who argue that nuclear power can simply be abandoned in favour of alternative energy sources and/or increased coal production. Nuclear power cannot be completely abandoned any more than coal can be discarded, whatever the disadvantages to health and the environment, not to mention security in the case of nuclear power. The complete abandonment of nuclear power would only become an available option if the whole of worldwide industrial society switched to a low-energy consumption life-style, in other words if an alternative culture were to be adopted by the next generation.

Meanwhile, the debate about the extent to which Britain should make itself dependent on nuclear energy continues to focus on whether Britain should maintain her present nuclear programme, reprocessing the waste at Windscale, Cumberland, into plutonium to fuel the next generation of fast-breeder reactors, to hold over the non-processed waste for up to fifty years, or to defer waste disposal for around ten years when a new, safer type of reprocessing plant might become available.

5.22 ALTERNATIVE ENERGY

While alternative energy sources such as wind, wave and solar energy will probably still need to be regarded as a minority source of supply right up until the year 2000 and beyond, their respective possibilities for faster exploitation should not be underestimated. In particular, a recent paper by the Astronomer Royal, Sir Martin Ryle, entitled *The Economics of Alternative Energy Sources* skilfully points out that some of the alternative renewable forms of energy may have a better prospect for early development than the government, heavily committed to, at worst, propping up the state energy sectors and, at best, in extracting a better performance record from the same traditional sources of energy, is prepared to admit.

The core of Sir Martin's argument hinges on the advocacy of wind power as a means of producing electricity more cheaply than any of the nuclear power alternatives so far envisaged. This arises in part from the nature of the demand.

A significant part of British consumption is for the heating of houses and other buildings; the demand for space heating, as it is called, shows large fluctuations between day and night, between summer and winter, etc, thus giving a peak much higher than the average. It thus comes down to a question of storage, whether a fuel for such a demand is expensive or inexpensive. At present, the replacement of oil and gas-fired space heating envisaged for the medium-term future requires the construction of a mammoth network of nuclear power stations to meet the peak loads at present supplied by oil or gas. If such a programme could be justified on a cost basis, it would remain questionable whether this is technically feasible in the time available. Furthermore, if the fastbreeder reactor were to be adopted as the basis for such a programme, a prototype would have to be built and tested before the main programme could proceed. Even the test programme would probably not be completed before 1987–9, which would put the main programme under an enormous time pressure.

However, the scope for renewable energy sources arises from the fact that the new capacity required can be reduced if arrangements can be made for storage of energy for a few days. This can be achieved by the storage of heat at the point of consumption, in homes and individual buildings, for example, as enlarged hot-water tanks. This greatly reduces the costs of the present systems which are based on a reconversion of electricity process. Such a storage system, or some variation of the same theme, eliminates at a stroke the most serious traditional objections to wind, wave and solar energy, namely their variability of output and transportation costs. In the case of wind power there are also environmental drawbacks. Moreover, both the capital and operating costs suggest that energy can be produced from wind for about one third the costs of a nuclear system. Wave power looks like being more costly than wind but there are possibilities for cost improvements. Finally, solar energy, which has such tremendous possibilities abroad, is the least promising of the renewable alternatives in Britain notably because of the unreliability of the sun's heat and the problems of storing what there is for when it is mainly required.

5.23 CONCLUDING PROPOSALS

In general this chapter has attempted to put forward proposals as the particular issues were discussed. However, if in this brief summary section there are some proposals that have not been previously aired, that should be the exception rather than the rule. The intention is to provide the briefest possible overall set of conclusions on the guiding principles which should underlie any attempt to open British energy resources up to the winds of competition, something which she is well able to do now that the principal energy sources, the principal fossil fuels, are already or about to become predominantly indigenous in origin. The proposals fall into six sections.

(1) *General*

In order to provide the framework for an orderly transition from an energy market comprising a host of miscellaneous restraints to competition to one where the fullest possible scope for competition reigns consistent with the interests of protecting those who most need protecting, this chapter recommends two major institutional changes. They are the creation of a reformed Energy Advisory Commission which is wholly independent of the Department of Energy and thus in a position to offer reasonably impartial advice; also a Select Parliamentary Committee on Energy to monitor the performance of the latter and the energy industry as a whole from the viewpoint of the individual citizen and the national interest.

(2) *Oil*

In order the better to provide for an efficient use of one of Britain's most valuable natural resources, the discretionary system of awarding licences should be abandoned in favour of an auction system which would simultaneously provide a higher economic rent for the nation and allow the most appropriately equipped companies to extract the oil in the fastest possible manner.

TABLE 5. *Britain, OECD Energy Balances, 1977*
Energy Balance Sheet (MTOE)

	Solid Fuels	Crude Oil & NGL	Petroleum Products	Gas	Nuclear Power	Hydro & Geothrm	Electricity	Total	P.E.E.E.*
United Kingdom 1976									
Indigenous Production	72.22	12.05	0.41	33.19	8.91	1.26	—	128.04	
Imports (+)	1.91	92.34	11.02	0.88			—	106.15	
Exports (−)	-1.43	-4.38	-16.15	—			-0.01	-21.97	
Marine Bunkers (−)	—	—	-3.51	—			—	-3.51	
Stock Change (+ or −)	-1.52	-0.90	0.57	—			—	-1.85	
Total Energy Requirements (TER)	71.18	99.11	-7.66	34.07	8.91	1.26	-0.01	206.86	
Statistical Difference	—	—	—	—			—	—	—
Electricity Generation	-44.35	—	-11.94	-1.79	-8.91	-1.26	23.82	-44.43	68.25
Gas Manufacture	-0.03	—	-0.27	0.12			-0.03	-0.21	0.09
Refineries	—	-99.06	91.65	—			-0.29	-7.70	0.83
Own Use by Energy Sector and Losses	-4.69	-0.05	-0.06	-1.73			-3.86	-10.27	11.06
Total Final									

of which						
Iron & Steel	6.60	2.73	1.00	1.13	11.46	3.24
Chemicals	0.13	2.36	4.82	1.68	8.99	4.81
Petrochemicals	—	—	—	—	—	—
Other Industry	4.09	14.27	7.05	5.04	30.45	14.44
Transportation	0.05	30.16	—	0.25	30.46	0.72
of which						
Road	—	23.76	—	—	23.76	—
Rail	0.04	0.90	—	0.25	1.19	0.72
Air	—	4.25	—	—	4.25	—
Navigation**	0.01	1.25	—	—	1.26	—
Other Sectors	11.24	11.67	17.80	11.53	52.24	33.03
of which						
Agriculture	0.04	1.33	—	0.31	1.68	0.89
Commercial Use	—	—	—	—	—	—
Public Service	—	—	—	—	—	—
Residential	11.20	10.34	17.80	11.22	50.56	32.14
Non-Energy Uses (Not included elsewhere)	—	10.65	17.80		10.65	
Electricity Generated — GWH		36154	5121	276976		
Efficiency in Electricity Gen. — percent		34.9	34.9	34.9		

Stock Drawdown +/Stock increase —

MTOE = Millions of Tons of Oil Equivalent

* — P.E.E.E. = Primary Energy Equivalent of Electricity

** — Internal and Coastal Navigation

(3) *Gas*

So that the gas price might be determined by real market forces rather than arbitrary, if well-meaning, political or even bureaucratic judgements, the gas monopoly should be pared by allowing gas to be sold directly by companies to major industrial consumers leaving the British Gas Corporation to look after the more variegated needs of household consumers. To promote less insulated policies, British Gas should be opened up to widespread share ownership by the public.

(4) *Coal*

In order to keep the degree of protection that the nationalised coal industry has enjoyed for so long within bounds, the legal restaints on competition should be lifted, allowing both cheaper oil or coal imports to come in and equally releasing the NCB from recurring interference with its prices and wages, etc. A future policy which would need to be agreed between the NCB and National Union of Miners in some detail before it was put to Parliament, would be to offer and encourage equity participation in the coal industry by individual miners who wished to take a stake in the future of their own industry under specially favourable terms.

(5) *Nuclear*

In order to maintain a reasonable balance between the requirements of the nation after oil and gas have begun to grow less plentiful and recalling the need to protect the environment as well as providing for national security, a limited expansion of Britain's nuclear programme should be approved, but with strict limits on plutonium expansion. The nuclear field does appear one of the sectors of energy supply where government regulation should be maintained very strictly. Nevertheless, in terms of investment, the relative merits of alternative energies should be continually reassessed so that they will be able to challenge nuclear power if it should appear that they can produce sufficient energy at the right price.

(6) *Alternatives*

In order that any one of them might be able to forge ahead rapidly if it once demonstrated its commercial superiority over its rivals, increased investment should be set aside for wind, wave, and solar power with the possible creation of an entirely new research institute to act as a focus for progress in these fields. The institution might become known as the International Institute for Alternative Energy Sources.

In conclusion, it may legitimately be noted, in the light of the evidence of this chapter, that the free enterprise system, far from being incompatible with a coherent national energy policy, as has been widely suggested, responds best when the political decision takers are following a consistent set of principles of which the most valuable is the principle of non-intervention unlesss absolutely unavoidable. The best qualified people to make decisions about energy are the consumers of energy — whether individual or corporate — for whom a competitive system is expressly designed.

6 United States Energy Policy

Throughout the 1970s, and especially since 1973, it has been patently apparent that a comprehensive turnaround in the pattern of US energy was essential to the future health of the world economy. An effective US energy policy has been eagerly and anxiously awaited by many governments for some time. After the false dawn of President Nixon's Project Independence, probably overly ambitious, certainly undermined by the faltering course of the Administration during Watergate, the problems of import dependence and demand exceeding supply began to accumulate and with it an ominous strengthening of the OPEC nations' ability to dictate terms not only to the United States but to most of the rest of the world.

Long before his election to the White House the aspiring Jimmy Carter recognised the high priority that energy would need to take, reserving the talented Dr Schlesinger for this demanding challenge. That within three months of his inauguration the new President was able to present a wide ranging programme is a pointer to the significance which he and most other members of his Cabinet attach to it. Such is the significance of the Carter-Schlesinger proposals for the worldwide community, not least Britain, that despite the fact that many of the proposals may never be enacted into legislation, they require the most careful scrutiny not only in terms of their possible effects on the US economy and other economies but also as a broad example of the range of possibilities that other Western societies might adopt and adapt to their own special circumstances.

In a special joint session of the Congress held on April 20 1977, the President declared that the energy problem constituted 'the greatest domestic challenge our nation will face in our lifetime'. The heart of the problem, warned President Carter, was that demand for fuel was rising more quickly than production. The answer was to reduce waste and inefficiency, he affirmed. Thus, in his very opening passage Mr Carter revealed conservation as constituting a major, if not the major, emphasis in his initial programme proposals.

6.1 THE PROBLEM

The President went on to outline the precise nature of the US energy problem: oil and natural gas made up 75 per cent of national consumption but they represented only about 7 per cent of US reserves. Demand for oil had been rising by more than 5 per cent annually but domestic oil production had been falling steadily by more than 6 per cent. Most seriously, imports of oil had risen sharply with the very real prospect that early in the 1980s even foreign oil might become scarce. If world demand for oil continued to rise during the 1980s at the current rate of 5 per cent annually, the proven reserves of oil could be exhausted by 1990.

Moreover, these fundamental energy dilemmas had the profoundest significance for the United States' position in the world, most obviously because it made her strategically vulnerable, but also because it fostered a growing trade deficit. During 1976 the United States imported more than 35 billion US dollars worth of oil (or around 40 per cent plus of her requirements). The need for a comprehensive energy plan, the President underlined, had long been overdue. If shirked, the consequences for the long term could be catastrophic.

6.2 GOALS FOR 1985

The President began the unveiling of his proposals by re-iterating the principles which he had first outlined on television earlier that week. They comprised seven specific goals:

(1) to reduce the annual growth rate in energy demand
   to less than 2 per cent;
(2) to reduce gasoline consumption by 10 per cent;
(3) to cut imports of foreign oil to 6 million barrels a day,
   less than half the level it would be if we did not conserve;
(4) to establish a strategic petroleum reserve of one billion
   barrels, about a ten months' supply;
(5) to increase coal production by more than two thirds,
   to over one billion tons a year;
(6) to insulate 90 per cent of American homes and all new
   buildings;
(7) to use solar energy in more than two and a half million
   homes.

In order to achieve those seven objectives the President suggested five central categories or means by which such goals
might be achieved, namely, (a) conservation, (b) production,
(c) conversion, (d) development and (e) fairness or balance.

6.3 CONSERVATION

The first and foremost goal of the Carter energy proposals
is conservation, not least because it is in the Administration's
view the cheapest, most practical way to meet energy needs
and to reduce growing dependence on foreign oil supplies.
The two main areas where the President pinpointed waste
taking place were transportation and heating and cooling systems.

Transportation consumes 26 per cent of US energy, of which
roughly half is waste. In Europe, for instance, the average
car weighs 2700 pounds; in the United States it weighs 4100
pounds. Congress, having previously adopted fuel efficiency standards requiring new cars to average 27.5 miles per gallon by
1985 (instead of the 18 they currently average), was requested
to adopt a graduated excise tax on new 'gas guzzlers' that
do not meet federal average mileage standards. The tax will
begin at a very low level and then rise each year until 1985.

In 1978, a tax of 180 dollars will be levied on a car getting 15 miles per gallon; by 1985 the comparable figure would be 1600 dollars for a car with the same mileage performance. All money levied from such a tax would be returned to consumers through rebates on cars that are more efficient than the mileage standard. The President expressed the belief that both efficiency and total automobile production and sales would increase under his proposals.

Among the most controversial proposals in the transport section of the President's conservation measures was the standby tax on gasoline, current gasoline consumption representing half of total oil usage. If implemented, such a tax would mean that between 1977 and 1980 gasoline consumption could be expected to rise only slightly above the present level. For the following five years, from 1980 to 1985, when more fuel efficient cars would have been brought into use, consumption would need to be reduced each year if the consumption targets are to be realised in 1985. In addition, to safeguard the effectiveness of his proposals, the President suggested a gasoline tax of an additional 5 cents per gallon be written into law that would automatically take effect every year that the United States failed to meet its annual target. Any proceeds from the tax, the President added, should be returned to the general public. As an important aspect of building equity into his programme, the President also advocated that the state governments be compensated for their loss of gasoline taxes by the creation of a Highway Trust Fund.

Heating and cooling systems, the other major area for conservation measures, means effectively the reduction of waste in homes and public buildings. Many buildings, the President pointed out, waste at least half the energy they use for heating and cooling. New buildings must be of improved efficiency and old buildings must be re-equipped or 'retrofitted' with heating systems and insulation that dramatically reduced the use of fuel.

The President proposed that the Federal Government should set an example. To this end the White House would issue an executive order establishing strict conservation goals for both new and old federal buildings — a 45 per cent increase in energy efficiency for new buildings, and a 20 per cent increase for existing buildings by 1985.

To help those who own homes and businesses to conserve the President proposed that those who weatherise buildings would be eligible for a tax credit of 25 per cent of the first 800 dollars invested in conservation, and 15 per cent of the next 1400 dollars. Other proposals for conservation in homes and buildings included (a) direct federal help for low-income residents, (b) an additional 10 per cent tax credit for business investments, (c) federal matching grants to non-profit schools and hospitals, as well as (d) public works money for weatherising state and local government buildings.

Responding to the waste encouraged by the utility rate structure which has tended to reward waste by offering the cheapest rates to the largest users, the President offered a package of proposals to be adopted over the next two years. They included the phasing out of promotional rates and other pricing systems that make natural gas and electricity artificially cheap for high volume users and which do not accurately reflect costs; offering users peak-load pricing techniques which set higher charges during the day when demand is great and lower charges when demand is small; individual meters for each apartment in new buildings instead of one master meter.

One final and far from insignificant step toward conservation proposed by the President was to encourage industries and utilities to expand 'cogeneration' projects, which capture much of the steam that is now wasted in generating electricity. In West Germany 29 per cent of total energy comes from cogeneration compared with only 4 per cent in the United States. To promote 'cogeneration' the President recommended a special 10 per cent tax credit for those who invested in it.

Taken together, the President's proposals for conservation represented the most far-reaching ever proposed by any US President and constituted perhaps the most important single emphasis in the energy programme so far unveiled. The conservation proposals were clearly aimed at arresting the relentless escalation of national energy consumption with consequences designed to strengthen not only the national strategic and economic interest but the future prospects of the entire Western industrial system, a fact gratefully acknowledged at the London economic summit of May 1977. Without much doubt the conservation proposals, if accepted by Congress, would become the most impressive attempt to tackle the waste of energy by any major country, a remarkable achievement in itself for a nation weaned on high consumption patterns from its inception.

6.4 PRODUCTION

The second major aspect of the President's proposals falls under the broad heading of production and 'rational pricing', as the President termed it.

In the President's view, US production of oil and natural gas cannot foreseeably be increased to meet US demand. However, the United States must be sure that its pricing system is sensible, discourages waste and encourages exploration and new production. As a means of bringing supply and demand into balance, over the long run, the price of energy should reflect its true replacement cost. This principle is particularly important for the nation's scarcest fuels, oil and natural gas. However, in the President's opinion, the pressure for immediate and total decontrol of domestic oil and natural gas prices would be disastrous for the US economy and for all Americans. It would notably not solve long range problems of dwindling supplies.

The President's central proposal on production was that the price of newly discovered oil would be allowed to rise, over a three-year period, to the 1977 world market price, with allowances for inflation. The current return to producers for previously discovered oil would remain the same, except for adjustments because of inflation. Moreover, in order to avoid windfall profits to the producers and yet at the same time to recognise the replacement costs of energy in the pricing system, the President proposed a wellhead tax on existing supplies of domestic oil equal to the difference between the present controlled price of oil and the world price, and return the money collected by this tax to the consumers and workers of the United States.

Alluding to the endemic interstate conflicts on oil and gas, which reached uncomfortable proportions during the shortages of the harsh winter of 1976–7, the President recommended a policy to end the artificial distortions in natural gas prices in different parts of the country which have caused people in the producing states to pay exhorbitant prices, while creating shortages, unemployment and economic stagnation, particularly in the north-eastern states. In the President's own words: 'We must not permit energy shortages to Balkanize our nation'.

The President pronounced his intention to work with Congress to give gas producers adequate incentive for exploration, working carefully toward deregulation of newly discovered natural gas as market conditions permit. Specifically the President proposed for the moment that the price limit for all new gas sold anywhere in the country be set at the price of the equivalent energy value of domestic crude oil, beginning in 1978. This proposal would apply to both new gas and to expiring intrastate contracts. It would not affect existing contracts. This section on production and pricing was perhaps fairly predictably the most heavily criticised by the parties most directly affected by such measures, the producers. Detailed criticism can be reserved for the latter part of this chapter.

6.5 CONVERSION

Following on naturally from the production of oil and gas was the President's third major plank in his programme, the conversion to coal from the former energy sources and in turn coal's conversion into more efficient sources of energy. It is imperative that oil and natural gas are not wasted by industries and utilities that could quite readily use coal instead, the President argued.

Although coal provides only 18 per cent of US energy requirements, it makes up 90 per cent of US energy reserves; while its efficient production and use create environmental problems, these can be met by strict strip-mining and clean-air standards respectively.

The President's targets for coal conversion were highly ambitious: namely to increase the use of coal by 400 million tons or 65 per cent in industries and utilities by 1985. To achieve this the President proposed a sliding scale tax, starting in 1979, on large industrial users of oil and natural gas. Utilities would not be subject to these taxes until 1983, because it would take them longer to convert to coal.

Also submitted to Congress were proposals for expanding research and development in coal. The fundamental quest was to find better ways to mine it safely, burn it cleanly and use it to produce other clean energy sources. The President remarked on the fact that billions had been spent on nuclear power but very little on coal.

Even with a successful conversion to coal programme, the President pointed out there would be a gap between the energy the nation needed and the energy that could be either produced or imported. This means that, to quote the President, 'as a last resort' the United States must continue to use increasing amounts of nuclear energy.

Currently the United States has 63 nuclear power plants producing about 3 per cent of their total energy and about 70 more are licensed for construction. Domestic uranium supplies can support this number of plants for another seventy-five years. Effective conservation efforts can minimise the shift toward nuclear power. As far as the United States is concerned, the President reiterated what he had only recently announced (a week earlier) namely, that there was no need to enter the plutonium age by licensing or building fast breeder reactors such as the proposed demonstration plant at Clinch River.

The President, however, did outline the direction and framework within which nuclear power plants might be expanded in numbers by proposing capacity be increased to produce enriched uranium for light water nuclear power plants, using the new centrifuge technology which consumes only about one tenth of the energy of existing gaseous diffusion plants. He also promised a reform of nuclear licensing procedures to avoid new plants being located near earthquake fault zones or near population centres; designs should be standardised as much as possible, and more adequate provision should be made for spent fuel storage.

6.6 DEVELOPMENT

The fourth element in the Carter strategy outlined to Congress was the development of permanent and reliable new energy sources. Almost certainly the most promising of new energy sources is solar energy. Much of the technology is already available, solar water heaters and space heaters are currently ready for commercialisation. The missing ingredient is some extra incentive to initiate the growth of a large market. Responding to this requirement, the President proposed a gradually decreasing tax credit, to run from 1977 until 1984, for those who purchase approved solar heating equipment. Initially, such credits would be 40 per cent of the first 1000 dollars and 25 per cent of the next 6400 dollars invested. Another important alternative energy source which the President pointed out he wished to be rapidly exploited was geothermal energy. This could be achieved, he suggested, by providing the same tax incentives as for gas and oil drilling operations.

The President repeatedly stressed the concern he had that the federal energy programme would above all be fair to all American citizens and interest groups. 'None of our people must make an unfair sacrifice. None should reap an unfair benefit', said the President, who claimed that the desire for equity was reflected throughout the entire plan, specifically in the wellhead tax, which encouraged conservation but was returned to the public — in a dollar-for-dollar refund of the wellhead tax as it affects home heating oil; in reducing the unfairness of natural gas pricing; in ensuring that homes will have the oil and natural gas they need, while industry turns toward the more abundant coal that can also suit its needs; in basing utility prices at true cost, so every user pays a fair share; in the automobile tax and rebate system, which rewards those who save our energy and penalises those who waste it.

6.7 BALANCE

The final plank in the Carter energy strategy falls under the category of balance. To achieve that logical and rational dispersion of effort and resources, the plan required the maximum of accurate information about supplies of energy and about the companies that produce it. As the President expressed it:

If we are asking sacrifices of ourselves, we need facts we can count on. We need an independent information system that will give us reliable data about energy reserves and production, emergency capabilities and financial data from the energy producers. I happen to believe in competition, and we don't have enough of it.

During this time of increasing scarcity, competition among

energy producers and distributors must be guaranteed. I recommend that individual accounting be required from energy companies, for production, refining, distribution and marketing — separately for domestic and foreign operations. Strict enforcement of the anti-trust laws can be based on this data, and may prevent the need for divestiture.

The President went on to urge that profiteering through tax shelters should be prevented, moreover that independent drillers should have the same intangible tax credits as the major corporations. The energy industry should not reap large un-earned profits. Increasing prices on existing inventories of oil should not result in windfall gains but should be captured for the people.

We must make it clear to everyone, that our people, through their government, will now be setting their energy policy. . . . The new department of energy should be established without delay. Continued fragmentation of government authority and responsibility for our nation's energy programme is dangerous and unnecessary.

6.8 DEMAND

The Carter proposals represent the first concerted attempt by the largest national consumer of world energy to reduce the rate of increase in energy demand. As such it must be welcomed whatever particular amendments may be necessary to make the Carter proposals both acceptable to Congress and workable in practice. In its scope and sweep in the area of conservation the Carter proposals deserve the highest praise. Not only do they represent a serious attempt to grapple with patterns of energy consumption which have grown up relentlessly since the colonial period but, in principle, if not in detail, they offer an example to other high energy consuming nations. For if the richest nation in the world is prepared to tackle energy waste head-on, with all the readjustments and political unpopu-larity that such an approach implies, how much more need those other Western industrial nations, less economically resil-ient, to embark on comparable national programmes. The fact that at the beginning of 1977 the United States imported 40 per cent of its crude oil requirements has been the principal factor that has enabled the OPEC cartel to tighten its grip over the western industrial system. While the Carter proposals are designed to reduce substantially US import dependence, its effects, if implemented, are likely to be gradual rather than spectacular. Indeed, the whole nature of the conservation proposals are gradualist and their eventual effectiveness will only be able to be assessed some years hence. There is, in fact, more than a suspicion that if the supply measures in the President's programme prove ineffective, then the consequences of the conservation proposals will serve to strengthen rather than to weaken the power of the OPEC states. Thus while the general tenor and range of the demand regulating proposals is to be welcomed, they cannot in practice be assessed without properly examining the supply related proposals.

6.9 SUPPLY

The passage devoted to production in the President's April 1977 address to Congress is the least satisfactory aspect of the Carter proposals. While there is a specific acknowledgement of the need to provide incentives to exploration and development for indigenous oil and natural gas, and the President's proposals do include suggestions that world oil prices should be paid to producers for all *new* finds and exploitations, this is not likely to have any medium-term effect on the level of domestic energy production for either oil or gas. For the present the full system of price controls will remain in force. This means that oil from Alaska's north slope will not immediately be encouraged through any loosening up of the network of price controls.

Unlike Britain, whose energy industries are dominated by public sector enterprises, the United States energy industries are dominated by corporations of one kind or another. The richest and most powerful of these corporations have traditionally extracted oil and gas. But they have for some time diversified into becoming energy majors intent on extracting and developing a wide number of energy sources such as coal, solar and geothermal energy as well as the more traditional oil and gas. The capital with which to develop these alternative energy sources has previously derived from the profits of oil and gas. The managerial, technical and entrepeneurial skills represented by the oil and gas corporations are, in addition, quite as essential to alternative energy development as are its capital resources. The question which poses itself very strongly arising from the Carter proposals is, if there is no substantial loosening of price controls, whether the energy majors will in practice embark on the path of rapid alternative energy development in the shape of coal, solar or geothermal energy.

If the strength of the Carter proposals is their emphasis on the urgency of improved and sometimes painful conservation measures, its weaknesses lies in its implicit faith in legislative means rather than the operation of the market. Thus the energy imbalance cannot be corrected except by a combination of voluntary restraint, by major shifts in relative prices, or by rationing. To bring about all or any one of these changes will invite substantial political unpopularity. The central question is whether the Congress, despite the obvious attempt to maintain an equitable sharing of burdens, can reasonably be expected to pass appropriate legislation. The answer must be that there must be serious doubts right across the board.

Although Professor Milton Friedman's reported concern that the array of planned energy price controls and taxes would greatly depress economic activity, while adding to inflation, is predictable, given his longstanding commitment to market principles, there is a strong lobby that subscribes to the fundamentalist market viewpoint that the lifting of all energy price controls would then allow free market forces to work fully to produce both lower consumption and increased incentives to producers. It is virtually accepted that the original Carter incentives for new oil and gas developments were inadequate. At worst, the Carter programme, had it been adopted unamended, might have succeeded in making energy more expensive, reduced production of oil and gas, and strengthened OPEC by creating an increased scarcity of supply in the United States. Since it is not yet clear what combination of measures will in fact be deployed, it is probably too early to make any final judgement on the total programme.

What can be said with confidence is that the President in no way exaggerated the dimension of the US energy dilemma.

For while the President's programme was chiefly concerned with the present position extending up until about 1985, a subsequent independent report published in May 1977 by the US Department of Commerce covering US energy up to the year 2000 presents an even more sobering picture of US oil production (together with imports) proving insufficient. The report tends to underline the fact that the President's proposals are in the nature of absolute minimum measures necessary to sustain industrial activity and confirms the urgency of the need to expand conservation proposals and to switch to coal and nuclear power.

The Commerce Department report, basing its calculations on a 2 per cent growth in annual energy consumption each year for the next twenty years was considered by many as too conservative. A 2 per cent rate, for instance, is in fact about half the rate of the last few years. But even with conservative estimates for energy demand, the report predicts a spectacular fall in the production of oil from 10 million barrels per day up to 1985 to a mere 6 million barrels per day in the year 2000. The most significant revelation of the report was that oil imports were likely to remain at something in excess of 45 per cent, even if the Carter proposals were fully implemented by Congress. The spectre of an enduring OPEC cartel was unmistakable.

6.10 GENERAL CLIMATE

By the late summer of 1977 the brand new Department of Energy came into being, bringing together in a 20,000 employee bureaucratic conglomerate, something like a dozen government agencies leaving the FERC, the Federal Energy Regulatory Commission, as the only body remaining outside the Department's jurisdiction. While the Department was expected to take some time before it became a cohesive force, its formation and mode of operation was of considerable concern to the energy industry (including production, processing, transportation and distribution) and most especially whether the DOE and FERC would create a regulatory atmosphere inimical to domestic energy investment. This concern was particularly acute in the sphere of federal policy toward off-shore oil and gas, still in its infancy.

For the moment the US oil industry is still heavily concentrated within the bounds of the continental United States and its off-shore limits to the ratio of about two to one, that is roughly 2000 out of 3000 rotary oil rigs in operation around the world exist in the United States. While marine activity has languished as a result of delays in federal offshore leasing, oil exploitation on land has recently expanded significantly. Whether the DOE and the Congressional legislation that it may be empowered to administer will reduce this expansion remains to be tested. An identifiable bottleneck which has already arisen affecting future US gas supplies, specifically Arctic continental gas, is the uncertainty surrounding the route of the projected pipeline, whether it should be the Alcan pipeline incurring an annual Canadian surcharge of 200 million dollars, or the Canadian proposed Mackenzie Valley pipeline, or the El Paso proposal for an all-American project. In this example it is the Canadian Government that has raised the uncertainty, but it may represent the shape of things to come under the US Department of Energy, that is, a more interventionist climate in which the government takes upon itself increasing responsibility in the energy sphere.

Meanwhile, if the prospects for governmental regulation look set for a period of heightened activity, there has been at the same time paradoxically a great improvement in the attitudes in official Washington toward the oil industry. Whereas in 1973 the results of a poll revealed the climate to be 'generally unfavourable' to the oil industry, by 1977 it had become either 'favourable' or 'neutral'. The concrete consequences of such a switch in official attitudes toward a major sector of the energy industry have been effectively to reduce the chances of divestiture, not to mention windfall profits, tax, etc. and to enlarge the possibilities of a more rational public debate on energy policy issues. The consequences of this change in official mood is still uncertain but it may be that it will lead to a more sympathetic hearing of the oil corporation interests in the passage of President Carter's proposals to regulate their activities. It certainly cannot be discounted in the calculations about the demand and supply of domestic supplies of oil and natural gas. Moreover, it encourages more open and vigorous support of the kind seen in August 1977, when the midwestern governors' conference appointed a lobby committee to influence senate votes in favour of oil and gas deregulation.

6.11 DOMESTIC DEMAND

So far we have described and analysed the scope and nature of the Carter proposals and the climate in which they will operate. It remains to examine the Presidential proposals as a whole both in terms of their fundamental diagnosis of not only US energy prospects but also US economic prospects, since the two can hardly be separated, and simply extrapolating past trends in energy demand and supply is not really a sufficiently sophisticated means of constructing so central a policy. There needs to be a distinction drawn at the very outset between the wisdom of particular goals such as increased and improved means of conservation, which are unexceptionable under almost any conceivable scenario, and measures which are aimed at enforcing new consumption patterns by regulatory means. It is the latter which require our most serious scrutiny.

The underlying premise of the Schlesinger energy package that oil and gas are not merely finite and non-renewable but may well run into serious scarcity if not actually run dry before the end of the twentieth century is almost irrefutable. It nevertheless rests upon the continuation of recent historical rates of consumption around the world remaining unchecked. Manifestly this is a false assumption to make, not least because the United States is herself by far the largest single consumer of oil and gas and her historic rate of fuel consumption has already showed a slowing down in the growth of total energy consumption.

This brings us to the central flaw in the Carter energy proposals, the apparent underestimation of recent favourable shifts in domestic energy demand, that is, since late 1973. By not recognising the extent to which there is potentially a 'natural' turn down in the previously escalating energy consumption trends, the Carter Administration may have needlessly locked itself into an energy framework which, while imposing a battery of legal restraints, will also choke off sources of domestic supply. The very fact that a relatively small band of energy experts has taken it upon itself to be responsible for such an all-encompassing programme is in itself ominous to those who have examined the direction and control of the energy market in other countries. It may yet prove to be the case that the unimpeachable arguments for restraint will so permeate the public consciousness that many of the more Draconian measures will prove superfluous.

To illustrate, the Administration's goal is to achieve an energy growth rate of 2 per cent, a figure which many observers believe will be achieved in any event without the necessity of controls which may inflict harmful wider consequences on the economy as a whole. As we noted earlier, the US Department of Commerce has calculated a 2 per cent growth rate in annual energy consumption each year for the next twenty years, and while this may be half the rate of the last few years, it is not unreasonable to assess the rate through the mid-1980s to be not very much in excess of 2 per cent, even without the Carter energy restraints.

6.12 NATIONAL ECONOMY

The principal reasons for such expectations are based on very broad arguments backed up by a considerable body of evidence, which it lies outside the scope of this book to deploy, suggesting that the US economy is probably not going to expand much faster than 3 per cent from now (1977) until 1985. This judgement rests heavily on the belief that inflation will remain to plague the American people at an uncomfortably high level having, as it were, become almost institutionalised by being a more or less constant factor for very nearly the last decade. The stiff new fuel taxes proposed by the President could even make the level of inflation more acute than it would otherwise become.

Such expectations, while running counter to recent experience, are in fact not without historical precedent to support them. If, for instance, we examine the twentieth century experience of industrial America, we conclude that a 3 per cent to 3.25 per cent growth is more 'normal' than the 4 per cent rate of the last twenty-five years which followed in the wake of the Great Depression and the Second World War.

There is in addition a real possibility of a further recession from the inflationary excesses of the last ten years. Yet even without further government intervention, total energy demand could rise at a rate considerably less than real GNP during the next decade regardless of whether the economy became recessionary or expansionist in its overall characteristics. Already, the United States is getting 4 per cent more real economic output per energy input now than immediately prior to the OPEC revolution. Moreover, because of the increasing importance of electricity in the economy, the fairly significant changes in fuel consumption have not been fully realised because of the substantial energy conversion losses from electricity.

6.13 END-USE ENERGY

It is in the area of the improvement of end-use energy that some of the most striking changes are already taking place. If such changes are to be properly exploited they need much wider recognition. They include at least four main distinct areas of improvement: (1) the industrial sector as a whole which has demonstrated a gradual improvement in energy efficiency; (2) a future slow down in the postwar household and commercial sector's spectacular growth; (3) a future improvement in energy efficiency in the transport sector by 1985; (4) while the prospects are for a dangerously wasteful switch to electric power by the year 2000 — from 26 per cent now to 50 per cent then — the intermediate-term growth (ie before massive nuclear capacity comes into production) will taper off.

In the vital industrial sector, total energy requirement is expected by some analysts to rise only half as fast as that sector's production of goods by 1985. It is not necessary to accept such a judgement totally; it may not be quite so drastic an energy cutback through greater efficiency, to perceive the general trend which is in fact in line with historical experience. Industry's energy inputs grew at 2.3 per cent average annual rate from the late-1940s to the early-1970s, while industrial output was climbing at a 4.6 per cent pace. Such long-term increases in energy efficiency reflects both changes in the composition of industrial activity as well as primary changes within certain energy-intensive businesses. A good example of the latter is the switch in steel from open-hearth furnaces to the basic oxygen process.

Since fuel costs are more likely than not to outrun inflation in general, even without the Carter proposals, the incentive to economise will be stronger than ever before. In the opinion of many commentators, however, it must be recognised that the strongest gains have already been recorded. Thus US industry managed the same amount of production in 1976 as in 1973 with 10 per cent less energy consumption.

6.14 POLITICAL COSTS

As has already been alluded to in the recently preceding section on the general economic climate, since 1973 there has been a greater public awareness of both the very high risks and enormous capital costs of the energy sector. To put it bluntly, there was an urgent need that it should be so if domestic energy supplies of oil and gas in particular were not to run out considerably ahead of time. But this is only half the story since there is still considerable ignorance as to the extent to which the political costs of the energy industry have escalated steeply recently, the full consequences of which are only dimly appreciated by the Federal Government, judging by their neglect of incentives to the oil and gas industry in the original Carter proposals. It is therefore of some importance to examine just where some of the money, especially the extra earned income, is going since it is, in the last analysis, of equal importance to ensure that capital as well as energy is used efficiently.

It is a truism that the petroleum industry is a capital intensive rather than a labour intensive business. Thus its direct people costs have traditionally been considered small relative to the total costs of doing business. However, by the same token, its indirect people costs are relatively large. In the end most capital costs emerge as people costs, so that when the petroleum industry spends vast sums of money for a wide range of capital equipment, it is in effect paying people costs indirectly. Moreover, the industry typically makes extensive use of service organisations on a contract basis, another form of paying people costs indirectly. In consequence, a great many external jobs arise out of the external activities of the petroleum industry. In their turn many, if not the majority, of these jobs hinge upon the prevailing government policy emphasis.

But if the petroleum industry's people costs have risen rapidly in recent years, its political costs are rising very much faster still. These political costs comprise taxes of a very wide variety, plus fees and bonuses, in fact all the charges that government manages to extract for the privilege of doing business. Naturally, these political costs must be recouped from the price that the consumer pays for petroleum products. The figures already available reveal the extent to which the political levy on petroleum companies has grown dramatically. In a recent survey of a group of petroleum companies representing more than half the worldwide industry the following overall picture emerged.

First, political costs very nearly trebled between 1972 and 1976; secondly, around 85 per cent of the increased revenue was absorbed by higher operating costs, 11 per cent for increased taxes, leaving only 4 per cent of the revenue gain available for retention as profit; thirdly, the companies paid more than twice as much in taxes as they earned in profit in spite of the fact that tax payments represented a minor proportion of the total political costs. By far the greatest proportion paid over to the government is included in the operating costs.

While the above figures are remarkable enough in themselves, a further breakdown of the government 'take' from the same companies over the identical four-year period is even more startling. Thus in 1972 the group of companies paid 19 per cent of their total revenue to government; by 1976 they were obliged to hand over virtually half of all the money they took in, or an increase of 554 per cent in the government 'take' in only four years. In other words, of the increase in company gross revenue during the period, two thirds was taken by government. The stage has long since been reached where the political costs have become the principal component of the total cost of doing business. Despite this sudden increase in the government 'take' there are ample signs that government intends to take more and that its appetite has been whetted.

Abroad, the governments of many of the producer countries want higher prices still for their petroleum. Such increases when they are introduced will represent the equivalent of a tax since, under the cartel currently operating, prices have no direct relationship to cost. At home in the United States the whole trend is toward additional taxes on crude oil and petroleum products. Bearing in mind the essential nature of petroleum under the present economic structure, there is some doubt about the extent to which new taxes, if imposed, will significantly alter consumption. As the price of cars, the depreciation, the insurance and the maintenance costs went up, the use of automobiles did not noticeably decrease. Moreover, the general resistance to overall higher taxation is rising to match the growth in the overall government 'take' in the national economy. In 1976, the American people paid an estimated 1.6 billion dollars a day to pay for the combined cost of federal, state and local government, 55 per cent more than in 1972. The cost of government has for some time being growing faster than the growth in the national economy which taxes on crude oil and refined petroleum products will do nothing to mitigate.

The burden of the argument against increasing the political costs of petroleum production by taxes on crude oil is that, for most consumers, it is a tax that they can scarcely avoid paying. It is arguable that such inescapable taxes may lead to higher pay demands, leading in turn to higher prices for goods and services which adds up to an inflationary impact even though the Carter energy administrators assure the American people that this will not occur.

Finally, on the subject of political costs, if there is an absolute requirement to utilise energy responsibly and more efficiently, much more efficiently than in the past, there is also an equal responsibility to deploy financial resources efficiently, which means the whole sphere of the political costs. There is always the underlying danger that the political levy will become so onerous that it will in time drive out the commercial incentive to find new supplies and to develop them. Psychologically it is by itself harmful to have such a high proportion of available funds syphoned off to pay what can be loosely categorised as non-productive political costs. Even with an effective conservation programme, the nation's requirements for petroleum is likely to grow for some time yet, if not so rapidly as in previous decades. It is of fundamental importance, therefore, that having examined in the earlier part of this chapter the general and particular aims of the Carter energy proposals and offered a preliminary critique, we make a closer inspection of the underlying problems which the Administration and Congress will have to face, not simply now but probably throughout the remainder of the President's term of office.

But first a preliminary look at the nature of the struggle over energy within the Congress, generally believed to be a litmus test of both Congressional and presidential leadership during the Ninety-fifth Congress and quite possibly into the Ninety-sixth.

6.15 CONGRESSIONAL PROSPECTS

The Carter-Schlesinger energy proposals, more sweeping and substantive than any previous energy package presented to Congress, were laid before the most heavily Democratic Congress since the mid-1960s when President Johnson was able to win approval for important civil rights and social programmes. The Carter proposals are likely to be much more difficult to enact, not least because they affect so many private interests as well as virtually every senator and representative in the country where regional needs and differences are so striking.

As Congress saw it in the spring of 1977, the heart of the principal aims of the Carter proposals was to make energy cost more so that consumers, on their own initiative, would use less of it. In order to soften the impact the proposals recommended the return of some of the higher energy taxes through lower federal taxes. Indeed, it is in the sphere of taxation that the most crucial struggles have been conducted, notably in the Senate Finance Committee and the House Ways and Means Committee. In the Senate the newly-created Energy Committee has also been playing an important part in the shaping of energy legislation. In the House the legislation was divided between four committees, namely Ways & Means, Banking, Interstate & Foreign Commerce and Interior and Science. The bills originally drafted by these panels were returned to the newly established Select Energy Committee to be merged into comprehensive legislation before going to the floor of the House. The Select Committee had only limited power to amend the work of the other committees.

It is important to remember that the principal elements

in the Carter proposals were not original in that virtually all of them had been discussed by the 93rd and 94th Congress. The chief difference lies in the fact that President Carter has drawn together the various disparate elements in a unified policy framework and urged its rapid endorsement as an essential national energy conservation programme. The major emphasis of both the Nixon and Ford energy strategies was to increase domestic energy supplies; the major emphasis of the Carter programme is conservation backed up by increased domestic supplies. The main attack of the Carter programme is therefore concentrating on pressuring and requiring Americans to make scarce supplies of oil and gas go further and work harder for their keep, so to speak. The main opposition across the board in Congress, as we have already noted in our earlier appraisal, was the disincentive to domestic suppliers of oil and gas. On more general economic grounds the feeling in Congress before the legislative struggle began in earnest was expressed in the fear that the programme could add 2 per cent to the inflation rate by 1979, take around 40 billion dollars out of the economy by the same year and would need to be compensated for by tax cuts of around 14 per cent across the board, of which the latter seemed, to say the least, fairly improbable. From the beginning it was clear that the programme would receive no easy passage despite a general fund of goodwill and agreement about its objectives. And so it was to prove.

By the autumn of 1977 the Carter proposals, especially the tax aspects designed to further conservation, had received some severe buffeting in the Senate Finance and Energy Committees. Senator Russell Long, Chairman of the Finance Committee, who represents the oil-producing state of Louisana, championed the many in Congress who believed the energy companies needed increased exploration and production incentives, moreover that taxes are an inefficient means of enforcing energy conservation. In pursuit of these beliefs the Finance Committee voted against the Administration in favour of allowing income tax on petrol, also against the President's crude oil equalisation tax, effectively slowing petrol price rises. The Senate Energy Committee confirmed the latter decision. The ultimate defeat of the President's proposals to maintain controls on domestic natural gas prices by a Senate overwhelmingly in favour of deregulation had now become apparent to everyone. As in the North Sea, the inescapable truth was dawning that controls of one kind or another constituted the major cause of insufficient domestic output of gas. The remedy for this, even if deregulation were introduced immediately, would not solve the expected gas supply shortage overnight. Indeed, the problem of short-term US gas supplies is a major question deserving of close inspection and placed in its international context.

6.16 LIQUEFIED NATURAL GAS

Natural gas is still the major source of domestic energy, providing one third of all primary energy produced domestically and meeting a half of the nation's industrial energy requirements. The decline in domestic production could be reversed by the 1980s, but only if much more generous economic incentives to expand the search for new gas reserves are provided. As in the southern North Sea offshore gas fields there is not much doubt that gas reserves exist but the artificial restraints on the commercial reward that the open market would confer has served as a more than adequate damper on both exploration and development. Meanwhile natural gas supplies have needed to be imported from Canada, the Alaskan north slope, by gas from coal liquefaction and petroleum feedstocks, and last but not least, from liquefied natural gas imports.

One of the principal goals of the Carter energy proposals was to achieve (by 1985) a reduction in the imports of oil to a projected 6 million barrels a day. But such is the critical nature of the US total energy supply problem that gas imports are likely to be expanded rather than limited as contemplated by the Ford Administration. The leading initial question is, where will such gas imports be coming from in the future? The transport of natural gas by pipeline is not feasible on an intercontinental scale so the prospect for the future expansion of US gas imports lies mostly in the form of liquefied natural gas, LNG, transported by giant ocean-going tankers. There are in all about seventy countries that have commercial reserves of natural gas, two-thirds of them in the non-Communist world and two thirds in non-industrial countries where there is very little domestic demand for gas (eg Nigeria, Iran, Indonesia, Trinidad, Ecuador and several Middle East states) resulting in something like 90 per cent of such gas being flared. However, to become viable there needs to be sufficient reserves to justify the creation and maintenance of an LNG project for twenty to twenty-five years. After taking account of the quantity of the reserves, also the local demand, the quantity of oil, the costs of tankers and pipelines, etc, there appear to be around twenty-four potential sources of LNG.

Meanwhile, though the less developed countries' consumption of natural gas is predicted to grow at twice the rate of other non-Communist states, it will still be the United States, Japan and Western Europe who will consume the lion's share — 85 per cent of the consumption of the non-Communist world in 1975 — and this proportion is likely to decrease only slowly. In Western Europe, natural gas, as we have seen in earlier chapters, is produced primarily from reserves in the Groningen fields in the Netherlands and parts of the North Sea, also a little from inland Germany, France and Italy. Nevertheless, by 1985 it is expected that one quarter of Western Europe's gas supply is likely to be imported from the Soviet Union, Iran and Algeria. Japan, already a major importer of gas, is likely to become more so since the public demand to combat pollution is calculated to expand the quantity of LNG imports very significantly.

Thus, to summarise the development of the international trade in LNG, the principal markets are the United States, Japan and Western Europe. The first regular commercial trade began in 1964 between Algeria and the United Kingdom and was followed the very next year by the opening up of regular LNG trade between Algeria and France. By the mid-1970s there were no less than six regular scheduled LNG trades to countries in Western Europe. By 1971 Japan had begun to import LNG from Alaska, the same year paradoxically that the United States began to import LNG from Algeria to Everett, Massachusetts, although at this time the United States was still a net exporter of LNG. Beginning in 1978, the United States will be receiving initial deliveries from a large scale LNG project from Algeria; by 1979 it is anticipated that the import flow will have reached 1 billion cubic feet per day, a level expected to be more or less maintained for the succeeding twenty-five years. By 1977 there were ten LNG trades in existence with a further seventeen due to come into operation by 1985.

In 1977 LNG imports to Western Europe and Japan were about equal, with the United States (due to her own exports) far behind. By 1980 they are expected to increase substantially to all three regions but most of all to the United States, making them about equal. But by 1985 the likelihood is that Japan will have significantly outstripped Western Europe in LNG imports. In turn, the United States is most likely to be importing nearly double the quantity imported by Japan. Altogether the quantity of LNG imports to these three regions is likely to multiply between 1977 and 1985 about six times over.

On the exporting side, the two dominant exporters for the short-term future are Algeria and Libya, whose position is likely to remain virtually unchallenged until the mid-1980s when Indonesia will become a principal supplier. By 1980, LNG trade will have reached about 5 billion cubic feet a day, at least 60 per cent of it being transported from Algeria to ports in the United States and Europe; the remaining 40 per cent drawn from around five countries will be destined for the Japanese market. By 1985 LNG imports may have reached 12.6 billion cubic feet, 50 per cent from Algeria still, with Indonesia having become the next most important supplier.

In projecting the commercial future of LNG, and its future international development in particular, the principal single investment remains the LNG cryogenic tankers. An American built tanker, for instance, might cost anything between 120 and 140 million dollars. At the moment, in very broad terms, LNG is on a par with the landed cost of imported oil. In Europe, where both Algeria and Libya lie relatively close at hand, LNG possesses an advantage over imported crude oil though the future supply of North Sea gas may have an important effect on its future price.

Having made a brief analysis of the likely future world market in LNG, it is still not possible to be too emphatic about the actual future of LNG in US energy supply for the next decade since it is greatly, if not largely, determined by both the availability of domestic gas supplies and the national policy toward imported LNG. As to the first, on the evidence of the earlier part of this chapter there may be a much greater attention to creating incentives for domestic gas exploration and development if the Senate has its way. As to the second, without minimising the importance of security of supply and the overall policy objective of moving toward the greatest possible degree of self-sufficiency in energy, in the next decade there seems to be a very large possibility, even probability, that LNG imports will quite simply be needed to keep the US industrial economy at work. Whether, given the undeniable uncertainties that enter into highly capital intensive international commercial projects of this kind, the risk capital will be forthcoming is not absolutely certain. But the alternatives, mostly substantially increasing domestic gas production, are not so susceptible to such rapid expansion, or at least that seems to be the general opinion on the evidence available. The obvious political dangers of relying so heavily on two North African states whose political alignment is more likely to be in conflict than sympathy with overall US foreign policy cannot be ignored. On the other hand, there are grounds for believing that the very size of the commercial benefits, more especially for the more populous Algeria, may guarantee a degree of political restraint.

6.17 A PERSPECTIVE APPRAISAL

Expressed in its most fundamental form, the United States has sought ever since 1973, at first piecemeal fashion then more comprehensively, to hammer out an energy programme for the very pressing reason that the nation was beginning to run out of the most readily available sources of energy, namely oil and natural gas. It was these two sources that so dramatically increased their share of the energy market between the end of World War Two and the mid-1970s from roughly 50 per cent to over 75 per cent, for the excellent reason that they were able to be delivered more cheaply than any other fuels. The official concensus, with some significant dissenting voices still to be heard, was that that era was now drawing inexorably to a close.

This is happening because the cheapest and most accessible sources of oil and natural gas have already been found and are at various stages of their productive life. Although various reserves remain in the ground, they will, so the argument runs, become progressively more expensive to locate and extract. This is certainly true of the offshore fields as the history of

the North Sea corroborates, but it is an argument that has
been exaggerated. The search for these remaining reserves will
nevertheless continue for a few years at an unpredictable rate
depending on the effectiveness of the incentives that are deployed. This book provides considerable evidence that would
suggest that the market incentive of reasonable profits is the
key; others favour an extension of government planning and
regulation of various kinds. What is certain is that the search
will be gradually abandoned as the share of oil and gas in
the US fuel economy declines as the end of the century
approaches. For while the President's 1977 energy programme
has set goals to be achieved by 1985, the lead times needed
to make any important changes in sources of supply make
1985 a very short-term perspective. The reality is that there
is an urgent necessity to plan in broad terms for a generation
ahead. On a global scale the concluding chapter will attempt
to assess what real choices exist, the outcome of current trends,
and how the mix between the various fuels might be adjusted
for the sake of posterity, or at least the next century.

Again considering the widest possible horizons, energy sources
must be broadly categorised as embracing non-renewable fuels
such as oil, gas, coal and uranium which form the basis of
modern industrial society's fuel supply, and renewable sources
such as solar, geothermal and hydropower, also tidal and windpower. While this chapter has necessarily to concentrate its
discussion on the non-renewable sources, chiefly because they
need to be managed better than they have hitherto now that
their scarcity is becoming more acute, as the concluding chapter
underlines, the attitude of the US government in particular
toward the renewable sources of energy is crucial to the future
well being of not only the next generation of Americans but
the next generation of mankind. While the renewable sources
of energy may still be a generation away from commercial
exploitation they will, it needs to be said, remain in exactly
that position permanently unless there is a comprehensive, heavily financed research and development programme. Meanwhile
it is not only oil, gas, coal and uranium that need careful
husbanding but also oil shale and tar sands, too, which, though
they lack the immediate economic potential to match their
technological feasibility, are fairly certain to come into the
energy market within the medium-term future. But it is this

economic factor that constitutes the chief determining factor in energy use, a truth underlined by the fact that worldwide oil resources account for only 4 per cent of non-renewable resources (of which Middle East oil constitutes 1 per cent). The simple logic of the situation is that that 1 per cent is the most accessible energy source economically. Moreover, while coal and uranium resources are deemed to be sufficient to last for many generations, the same cannot be said about oil and natural gas. At recent historical growth rates oil production in the non-Communist world (based on existing reserves) will peak in about five years. But even supposing new reserves are discovered at the rate of about 20 billion barrels a year, production will peak in ten years. The growth rate of world oil consumption is clearly going to slow down. The United States has an additional, very basic problem.

The stark nature of that problem centres around the fact that domestic energy production had in 1977 reached a standstill. Thus for the immediate future any increase in energy use must be met from imports, principally oil imports that rose from 28 per cent in 1972 to 50 per cent in 1977. The US energy problem is thus essentially an oil supply problem in its most immediate ramification though it becomes obviously much more complex over the longer term.

Focussing on the oil problem, if we assume that the 1985 oil demand can be held to the present level, the goal of reducing oil imports to 6 million barrels a day will only be possible if domestic production increases by about 2.5 million barrels a day. By 1985, production from the North Slope of Alaska should provide about 2 million barrels a day, leaving about an additional half million barrels to be found from the rest of the nation's oilfields. Since production from existing fields is declining steadily, and by 1985 is expected to reach around half the present level, this could be difficult. Thus, the gap to be filled by production from reserves yet to be discovered, or developed through additional recovery of known reserves, is about 4.5 million barrels a day. Similar problems affect the supply of natural gas.

Natural gas production from existing fields in the United States has been declining steadily during the mid-1970s and if the incentive and opportunity to invest are not present then production in 1985 will fall below the present level. As things stand at the moment, throughout this period domestic gas production will be supplemented by modest scale imports by pipeline from Canada and in liquid form, as we have already examined in detail, from several overseas countries, especially Algeria and Libya.

It is apparent to everyone now that any increase in US energy consumption will have to rely in the short to medium term on coal or nuclear electric power. The only practical alternative to these two would be to allow higher oil imports. This central choice is recognised by the Carter proposals which support existing nuclear electric power plans as well as a stepped up emphasis on increasing coal production. The Administration's goal is to produce 1 billion tons of coal by 1985 — 400 million tons more than the current level. Even if it could be assumed that the environmental restraints on mining and burning coal could be successfully overcome, it would be a very substantial achievement if the targets for coal could be achieved in practice. Significantly for prospective West European importers of coal, there is no allowance made for US coal exports, even if the most ambitious domestic coal production target were fulfilled.

In general terms, as we noted earlier in this chapter, the Carter proposals seem well conceived in seeking to raise crude oil and natural gas prices so that US energy prices come closer to both the world market price and the cost of replacement. If spread evenly, the cost of oil products to consumers would increase by 7 cents a gallon. As instruments of demand management, the proposals are commendable, but as instruments of supply management, which is what is chiefly needed here, they are far from being the best instrument available.

The concept of increasing oil prices through taxation for one is largely irrelevant to the question of supply though it has certain advantages in terms of predictability for the government and private citizen alike. Under the Carter proposals several categories of oil production are defined, each with its own price and tax provisions. This has the obvious advantage of allowing the investor to assess with some degree of confidence whether it is better to invest in the oil production business or some other sector of the economy. However, on the current evidence he may well decide it does not.

The effective alternative to limiting consumption and improving the supply of oil and gas remains what it has always been — a simple acceptance of the virtues of deregulation of both oil and gas with safeguards such as windfall profit taxes to provide a just framework. For the inescapable evidence is that deregulation of oil and gas prices, while in no way guaranteeing an adequate return on new investments, still offers investors the hope of making a return commensurate with the risk and as such is historically infinitely more likely to stimulate investment.

But there are further weaknesses in deploying the tax system, and a fairly cumbersome set of taxes at that, to induce results which the application of market principles would produce much more effectively and more justly. Returning the various taxes that form part of the original Carter energy programme to the public as rebates will at best lessen the conservation incentive of higher prices. As a means of income redistribution it can only be described as arbitrary as any regional or occupational survey would starkly highlight.

But possibly the worst feature of the original energy proposals drawn up by the Carter Administration is that the rebated taxes will have a strong tendency to be turned into general consumption rather than turned over into investment in the energy industry. The evidence which has been outlined in this chapter has hinted at the growth in the government take and the degree of regulation compared with even five years ago. In practical terms this can only have a long-term disincentive effect on the energy industry which alone possesses the technical skills, the managerial know-how and has proved itself over several decades the best conduit of capital on a worldwide front. The truth is that if the incentives and opportunities to invest are provided, the petroleum industry will continue the search. And even though the reserves are finite, production can be substantially increased before the limits are reached. But if the regulatory climate for the industry becomes too oppressive (as distinct from plain interventionism as under the British system, if in general terms the industry is not sufficiently encouraged by government, then quite simply the era of oil and gas will end that much sooner than it needs to end. Without doubt that is in nobody's interest and must be avoided at all costs.

6.18 INTERNATIONAL PRIORITIES

Having made a general overall appraisal of the trends and
real requirements of US energy policy it becomes necessary,
as with any major national energy policy, to return to the
pressures provided by the international environment, a sphere
where political calculation cannot be divorced from economic
forecasting. Before making an analysis of the immediate prospects for the survival of OPEC and its current pattern of
price levels and adjustments, it is necessary to record an important agreement between the main consuming countries of the
non-Communist world which reveals their common appraisal
of the situation, its seriousness and the generally approved strategy to meet the challenge of an orderly world energy market.

The agreement in question announced in October 1977 was
nicknamed the Twelve Commandments and consisted of a series
of basic principles, agreed by the nineteen member countries
of the International Energy Agency, designed quite simply to
avert the possibility of any future collapse of the Western industrial system. Whether these guidelines are able to be followed
successfully over the next decade remains to be seen but there
now exists a written agreement among the member countries
of the IEA to restrict oil imports if possible to 26 million
barrels a day by 1985. This figure compares with 23 million
barrels a day in 1977 and a minimum of 42 million barrels
a day if present economic growth is maintained.

In the words of the US Energy Secretary, Dr James Schlesinger, if the measures drawn up by the IEA do not prove
effective, 'we would face social and political tensions not experienced since the thirties'. He added that the economic consequences of an energy failure would 'shake the foundations of
our society. We cannot afford to fail.' The two most single
influential components in such a co-ordinated programme are
clearly the United States and the European Communities, but
especially the United States. Dr Schlesinger promised that his
government would seek to limit oil imports to 5.8 million barrels
a day by 1985, compared with 8 million in 1977 and 16 million
as the projected need. The European Communities, for their
part, announced their commitment to try to hold their imports
to the 1977 level of 10 million barrels a day. The remaining
countries of the IEA are committed to limit their imports to

10 million barrels a day altogether. The agreement underlines dramatically the crucial nature of the United States and West European energy policies whose success or failure will greatly affect each of the other IEA member states. Above all, the United States, because of its stake not only in the world energy market but its enormous financial resources, has by the power of its example the most crucial role of all to play. Nobody in the Carter Administration is likely to forget that energy policy will probably make or break their government in the history books.

Deeply conscious of all these truths, President Carter followed up his initial effort in launching his energy programme in April 1977 with a further personal foray the same October, by which time his proposals had been cruelly mangled by Congress and most particularly by the Senate. The President accused the oil companies of seeking 'the biggest rip-off in history', added that his proposals would give US oil companies the highest prices for oil in the world and that deregulation (voted by the Senate against the President's wishes) would give the oil companies a price for gas fifteen times higher than three years previously. The reply was not long in coming with Mobil describing the President's remarks on the profitability of the oil industry as 'misleading'. It supported its defence by noting that the profits of the industry had fallen from 6.7 per cent of revenues in 1973 to 3.4 per cent in 1976 and pointed out that a Federal Energy Administration report published in June, 1977 concluded that the industry's profits were no greater than those of other manufacturing industries.

Whatever the merits of the respective arguments, what was indisputable was that in 1977 America imported very nearly half her oil at an annual rate of 45 billion dollars, which constituted the principal cause for its rapidly increasing and record high balance of payments deficit. The size of this deficit in turn was causing great concern in the currency markets, contributing to a decline in the value of the dollar. This alone would suggest pressure for further OPEC price rises, quite distinct from those arising from the general level of world inflation.

Nevertheless, by late-1977 the US Administration — the Department of Energy confidently, the Treasury less confidently — was predicting US oil imports would be cut in 1978 from a record level of 8.5 million barrels a day to 8.1 million barrels

TABLE 6. *United States, OECD Energy Balances, 1977*
Energy Balance Sheet (MTOE)

	Solid Fuels	Crude Oil & NGL	Petroleum Products	Gas	Nuclear Power	Hydro & Geothrm	Electricity	Total	P.E.E.E.*
USA 1975									
Indigenous Production	374.49	467.75		497.99	42.80	72.93		1455.96	
Imports (+)	1.66	202.01	93.86	22.74			0.97	321.23	
Exports (−)	−39.57	−0.28	−9.40	−1.78			−0.44	−51.47	
Marine Bunkers (−)		−	−18.15					−18.15	
Stock Change (+ or −)	−14.47	−1.50	−0.03	−1.42				−17.43	
Total Energy Requirements (TER)	322.11	667.97	66.27	517.53	42.80	72.93	0.53	1690.15	
Statistical Difference	0.05	−33.95	35.12	−				1.21	−
Electricity Generation	−234.41	−0.82	−76.91	−71.91	−42.80	−72.93	182.40	−317.39	499.78
Gas Manufacture	−0.20		−0.92	0.54				−0.58	−
Refineries		−633.19	613.60	−23.50				−43.09	−
Own Use by Energy Sector and Losses	−17.59		4.99	−44.72			−27.20	−84.52	74.32
Total Final Consumption (TFC)	60.96		642.14	377.80			155.73	1245.73	425.46

of which						
Iron & Steel	25.39	5.22	14.79	6.27	51.66	17.13
Chemical	0.10	—	0.62	11.97	12.69	32.70
Petrochemical	—	38.59	—	—	38.59	—
Other Industry	39.93	22.90	206.09	43.23	312.15	118.12
Transportation	0.01	409.09	—	0.38	409.49	1.03
of which						
Road		345.43			345.43	
Rail	0.01	13.10		0.38	13.50	1.03
Air		50.56			50.56	
Navigation**		—			—	
Other Sectors	4.53	121.70	156.39	93.87	376.50	256.47
of which						
Agriculture				1.32		3.60
Commercial Use			51.79	35.15		96.03
Public Service				—		—
Residential	4.53	121.70	104.60	57.41	156.84	156.84
Non-Energy Uses (Not included elsewhere)		39.66			39.66	
Electricity Generated – GWH			181645	309505	2120880	
Efficiency in Electricity Gen. – percent		36.5	36.5	36.5	36.5	

Stock Drawdown +/Stock Increase –
MTOE = Millions of Tons of Oil Equivalent

* — P.E.E. = Primary Energy Equivalent of Electricity
** — Internal and Coastal Navigation

a day. This expectation rested heavily on increased production from Alaska offsetting any rise in demand resulting from the Government's effort to build up a strategic petroleum reserve of 250 million barrels. The Carter energy objectives for oil imports in 1978 were based on an expected growth rate of 5 per cent in the economy as a whole. Since many economic forecasters saw such a growth rate as being over-optimistic by at least 0.5 per cent, if not 1 per cent, the expected lower levels of oil imports become that much more probable.

While from a strategic point of view the very high level of US oil imports represented a disturbingly exposed posture for a superpower, it was not so serious a matter in the purely economic perspective. Not only was the mammoth trade deficit of 1977 expected to be reduced in 1978, not least because of the projected reduction in the Japanese current account surplus with the United States, but also because in general the leading members of OPEC perceived it to be in their medium and long-term interests to anticipate future oil supply insufficieny. In any event the nearly one hundred year period, from 1893 to 1971, when the US sustained an automatic surplus with the rest of the world, was conclusively ended. Any attempt rapidly to reduce the current account deficit would probably merely induce a world recession. Moreover the US government perceived inflation to be a more potent and destructive enemy to be defeated than the current levels of unemployment, which would ultimately be perpetuated and enlarged if inflation were allowed to dig its toes in for some future economic debacle. In all this the roots of inflation were no longer laid at the doors of food, fuel and raw material imports as such but, increasingly, seen to lie in energy policy itself which allowed costs to rise and continued to impose a variety of taxes on the energy industry.

Inflation in the US economy clearly was due to a complex of ongoing factors both domestic and international in character but more and more energy policy was being monitored for its possible inflationary implications. If the Carter energy policy were eventually to turn out much more bland than its architects had intended, if the combination of powerful interests in the Congress could fairly readily be identified as blocking the President's original radical programme, the real victor over the Carter policy was the genie of inflation. It might be no more than a spectre by comparative international standards but its sting was known and feared by a combination of powerful economic interests whose perceived vital interests easily overrode the more generalised goals of the White House in the shape of conservationist, environmentalist and strategic objectives.

By the second half of 1978, on his return from the Bonn summit, President Carter had become deeply committed to a policy of oil equalisation designed to raise prices paid for US domestic crude (by means of a well-head tax) to the world level by 1980. His promise at Bonn — to have a completed energy programme enacted by the end of the year which would reduce oil imports by about 2.5 million barrels a day by 1985 — was viewed with scepticism in Congress. For while four-fifths of the President's energy programme looked like being enacted in a distinctly modified form the possibility remained that the President would need to take Administrative action to pass the final fifth of the energy Bill, that is the part pertaining to the raising of the price of domestic crude. The probability remains that the President faces a struggle extending into 1979 before his energy proposals gain the final endorsement of a jealous legislature. By and large the Congress approves of the President's goals, but believes his mode of achievement could be greatly enhanced by allowing the market to determine many of the issues without recourse to further regulation. The instincts of a substantial body of Congress, and especially of the Senate, accord fairly closely with both the arguments and evidence of this book.

7 The Prospects for Global Energy, 1985–2000

In treating the issues of national energy policies as we have done in the body of this book we have to a considerable extent concentrated on the near future, that is, the period up until 1985 or thereabouts. Yet the practical horizons of most governments and businesses are often of the order of five years rather than ten and the individual citizen is attuned to an even shorter perspective. The insidious danger arising from these varying levels of planning horizons is that the world energy supply situation could reach a critical imbalance in energy sources where serious economic distortions had been institutionalised by virtue of insufficient long-term planning. It is to minimise that propensity that this final chapter embarks on a very preliminary assessment of global energy factors and likely trends in the post-1985 period.

If one is to focus the nature of the global energy challenge before embarking on a more detailed analysis, it revolves around the central proposition that global oil production is likely to level off beginning in the mid to late-1980s. Since energy demand is still growing, if more slowly than previously, there is a need to produce several alternative fuels. However since such large investments and long lead times are involved if the prospective shortage is to be filled, there is a greater degree of urgency than might normally be associated with long-range planning exercises, more particularly since some of the options are likely to be keenly disputed. If, then, one were to summarise in a sentence what lies at the heart of the challenge to global energy analysts and planners it is no more and no less than to manage the transition from dependence on a predominantly oil economy (ie in the industrialised world) to a much greater reliance on other fossil fuels, on nuclear energy and on renewable energy systems. When we have finished examining what are generally regarded as the conventional range of options in the global energy market we can afford to make a more radical appraisal of the long-term possibilities for the most influential national energy market, that of the United States.

Quite the best summary of the future possibilities for the global energy market is to be found in the section on global energy futures contained in the WAES study, *Energy: Global Prospects, 1985–2000*, published in 1977. This contains a remarkably concise overview whose conclusions are possibly the best starting point for any examination such as this whose chief aim is to focus on what are the most important questions rather than argue each issue as in some of the national energy policy chapters. These conclusions fall under ten heads.

(1) The supply of oil will fail to meet an increasing demand before the year 2000, most probably sometime between 1985–95. This could happen even if energy prices rise 50 per cent above current levels in real terms.

(2) Demand for energy will continue to grow, even if governments adopt vigorous policies to conserve energy. This growth must increasingly be satisfied by energy resources other than oil, which will be progressively reserved for uses that only oil can satisfy.

(3) The continued growth of energy demand requires that energy resources be developed with the utmost vigour. The change from a world economy dominated by oil must start now. The alternatives require five to fifteen years to develop and the need for replacement fuels will increase rapidly as the last decade of the century is approached.

(4) Electricity from nuclear power is capable of making an important contribution to the global energy supply, although far from being universally accepted on environmental, political or economic grounds (see later argu-

ments in this chapter). Fusion power is not expected to be a significant source of energy before the year 2000.

(5) Coal has the potential to contribute substantially to future energy supplies. Coal reserves are abundant but taking advantage of them requires an active programme of development by both producers and consumers.

(6) Natural gas reserves are large enough to meet projected demand provided the incentives are sufficient to encourage the development of costly and extensive intercontinental gas transportation systems (see previous chapter).

(7) Although the resource base of other fossil fuels such as oil sands, heavy oil, and oil shale, is very large, it is likely to supply only small amounts of energy before the year 2000.

(8) Other than hydro-electric power, renewable sources of energy — solar, wind power, wave power — are unlikely to contribute significantly to additional energy during this century at the global level, though they could be very important regionally. By the twenty-first century they will almost certainly have become of increasing importance. As argued later in this chapter, they could be given such a high priority that they would in practical terms come to the fore even earlier, but this presupposes a fairly radical change of priorities and direction by the major industrial societies, beginning with the United States.

(9) Energy efficiency improvements (beyond the substantial energy conservation assumptions already built into our analysis) can further reduce energy demand and narrow the prospective gaps between energy demand and supply. Policies for achieving energy conservation should continue to be key elements of all future energy strategies.

(10) The critical interdependence of nations in the energy field requires an unprecedented degree of international collaboration in the future. In addition it requires the will to mobilise finance, labour, research, and ingenuity with a common purpose rarely attained in peacetime. If these conclusions are substantially right in the general

diagnosis, then failure to act in concert internationally could lead to (a) substantially higher energy prices, depressing the world economy, (b) the inevitable frustration of the growth expectations of the less developed countries and (c) political and social conditions generally which are likely to prove a focus for confrontation and conflict. Such is the considered judgement of the WAES report which represents one of the best orchestrated and researched projects into our energy futures so far conducted.

7.1 THE OIL ERA

It is now generally agreed that somewhere beyond 1985 global oil supply will plateau, marking the end of a fairly steady ascent, if one discounts the special circumstances of 1973–4. The remarkable feature of the twentieth century growth of industrialism is that it has been fuelled substantially on fossil fuels among which oil has provided the greatest share of the sharply increased overall consumption. While the 1950s and 1960s are traditionally regarded as the high tide of the oil era in terms of its competitive supremacy over other fuels, the likelihood is that oil production will show a striking growth from around 45 million barrels a day in 1973 to somewhere of the order of 70 million barrels a day by the late-1980s. As a very general statement it is a fairly safe forecast that energy demand will continue to grow simultaneously with economic growth, if more slowly than in the recent past. This slowdown in the rate of energy consumption growth is likely to arise out of a combination of saturation effects, more effective energy use in both the domestic and industrial sectors and not least the vigorous holding down of prices. It is interesting to note how since the early-1970s conservation, which had previously been a notion to which not much more than lip service was paid by both governments and businesses, has become a central strand in attempts to come to terms with energy futures.

Not only does the cumulative evidence point to the fact that a great deal more energy can very readily be saved with some forethought and planning but, just as importantly, that energy conservation may well be the very best alternative available, economically speaking. But while its advantages and benefits can sometimes be substantial, the rules for conservation cannot be laid down on a universalist basis but must be closely designed to fit the actual circumstances that prevail at a given moment in a given locality of the world. Two primary principles, sometimes placed into a position of competition with each other, are that conservation is best achieved from improved efficiency subsequent upon an increase in prices; the other principle is that they can be achieved from the application of selective government action through regulation, taxes and a variety of progressive restraints and incentives.

Many national energy planners are predicting improvements of around 1 per cent annually in the ratio of energy use to value added in industrial production to the year 2000. Some of this improvement in industrial energy use can also be reinforced by conservation in buildings either by improved insulation or improved combustion and control. In some of the colder advanced industrial nations such as Scandinavia, the United States and Canada, the savings in some regions can be as high as 40 per cent. Transport is another sector where striking improvements in efficiency can and are expected to be achieved both in miles per gallon and also the load efficiency. However, since demand for air travel is likely to increase fairly dramatically, transport could still prove to be among the fastest growing energy consumption sectors.

As President Carter among the major Western leaders, has most notably realised, conservation is the foundation upon which an effective long-term energy policy has to be built, more particularly for a nation so traditionally extravagant in energy consumption as the United States. In the adoption of a particular emphasis in a national energy policy there needs to be a fundamental distinction made between those policies which involve adaptation in order to improve efficiency and usually less fundamental changes which reduce consumption by bringing down the level of economic activity. The first sort of policy is normally planned in advance, the second most frequently adopted by governments from sheer necessity under the pressure of inescapable events. Most policies adopted by national governments represent an amalgam of the two sorts of policies.

In the end, while most energy conservation policies employ the means available to governments such as fiscal measures, government regulations and standards (supported in industry by such appointments as energy managers), there are around four main factors in almost every national energy conservation programme. They include the energy price, government action, structural changes in the economy and, not least, the climate of opinion expressed through the individual decisions of millions of users of energy. The last factor applies the more forcibly in a free enterprise economy.

What cannot be underlined too much is that because they invariably entail an extended and often complex process of decentralization, conservation policies take some years to prove their effectiveness, that is, their lead times are long — to use the jargon of energy planners. Thus, to take the most obvious examples, it takes a decade to change over a stock of cars and from twenty to thirty years for a comprehensive turnover of industrial equipment. At the extreme it typically takes a hundred years at the very least to change over a Western nation's housing stock. Essential to the effectiveness of a long-term energy strategy, conservation is now becoming a central feature not simply to the planning of the individual nation states but also the global energy strategies which provide the general guidelines within which national policies can best take root.

7.2 COAL

As the oil era gradually subsides it is not only conservationist policies that will be needed but substitute fuels on a hitherto undreamt of scale. There is not much doubt that coal has the potential to be one of the major energy sources in the period from 1985 onwards. Because of the size of the resource base, that is, the global reserves of coal, there is at least a theoretical possibility and practical probability that coal output will be significantly increased. What remains in doubt among the countries of the non-Communist world is whether in practice these countries will deem it wisest to make the required financial commitment. Whether they will do so depends almost entirely on whether they calculate there will be sufficient consumers for such a pronounced increase in coal production. There are grounds for both hope and doubt.

Part of the decline in popularity of coal during the oil era has not only been a question of price but also the fact that compared with oil, gas and electricity, coal is both dirty and awkward to distribute. Thus a feature of industrialised countries has been that coal has declined in its share of the energy market as both industrial and private consumers switched to more convenient and cleaner fuels. In many Western industrial countries coal is being increasingly confined to power stations and to the steel industry of which the latter is currently in deep recession, making coal production that much less profitable than it would otherwise be, but also exposing its essential economic vulnerability in times of economic recession.

The inevitable question whether this decline can be arrested cannot adequately be answered on a global basis since in some regions the grounds for optimism are substantial, in other regions virtually non-existent unless one is to embark on policies of permanent employment protection. Most of the longer term possibilities arise from the opportunities for technical improvements. Among the three main technical innovations utilising the present centralised energy system are coal's substitution for oil and gas in electricity generation, for process heating in industry and lastly coal conversion into oil and gas. The last means is especially expensive in both financial terms and also in terms of conservation since there are significant losses from the conversion process. In all three of these processes the costs and lead times for building the conversion plants, changing user equipment and fuel preferences is likely to be substantial.

From the resource angle, coal is beyond dispute among the most abundant global sources of energy in existence. The estimated economically recoverable proved global reserves of coal stand at 700 billion tonnes, which is the equivalent of 3000 billion barrels of oil. The potential reserves are even greater, possibly as high as 12,000 billion barrels of oil equivalent. These reserves, however, are very unevenly distributed. The so-called Big Three, that is the United States, the Soviet Union and China currently account for 60 per cent of all coal produced, with Poland, West Germany and Britain together accounting for another 15 per cent. For the immediate future, however, one of the most important areas for increased coal production lies among the Southern Hemisphere countries, of which the most notable example is Australia. The significance of coal to the less developed countries, both as a substitute for oil imports and a rich potential source of future exports, has not yet been fully appreciated.

Thus to summarise the global coal situation, there are beyond question abundant reserves available in many countries around the world. Given the rate of rundown of oil and gas there are some observers who believe the question which is most pressing is whether coal can be produced soon enough. This author believes that if the commercial viability were more consistently in prospect this would not present a problem despite the long lead times and enormous financial investment. At the moment that viability varies widely from country to country as the account in the body of this book would indicate.

If we extrapolate present trends and make various adjustments we might find that by the year 2000, something of the order of 3.2 billion tonnes might be produced annually of which the greater part by far would derive from North America, roughly 1.8 billion tonnes has been calculated which would represent a threefold increase in 25 years, a fairly astounding performance if it were in fact achieved. In order to do so coal's price in the market after providing for the cost of new transportation, coal handling and pollution control facilities would need to be sufficiently below that of competing fuels to encourage its use. Major technical improvements, as we shall see later, may yet give coal a promising future.

7.3 NATURAL GAS

Not unlike coal reserves, the global reserves of natural gas are fairly abundant and suggest that there is very little likelihood of production being limited during the next twenty-five years. However, there are serious prospective limitations in the sphere of transporting and distributing natural gas both physically from the well-head to the consumer and politically in the attitude of some natural gas producers toward the possibility of export. Since historically natural gas has been moved by pipeline it has until recently proved much too expensive to trade in internationally with something like 30 per cent of domestic total energy supply being met by natural gas in the United States and around 47 per cent in the Netherlands.

In order to exploit the global reserves of natural gas, it is quite clear that it must by some means or another be elevated from the role of a purely or largely domestic energy resource to a major commodity in international trade. To achieve such an objective a means of bulk transportation of natural gas has to be created which will be able to provide gas to the consumer at competitive prices. At the moment there are two principal alternatives. The first of these is liquefied natural gas (LNG) which requires refrigerated super-tankers to be built at great expense and also involves regasifying at the terminal centres. Not least of its disadvantages is that LNG involves an energy loss of something round 25 per cent. Despite these drawbacks there has already been a significant start made in the international trade in LNG as the chapter on United States energy policy documents. The second major alternative means of transporting natural gas internationally is to convert it into methanol. While this means is less expensive than LNG, it has the serious disadvantage, some would say fatal flaw, of involving a 40 per cent energy loss. Moreover, it has for the present a distance restriction of around 10,000 kilometres which is a serious liability considering the fact that the most important gas reserves exist great distances from the principal potential markets of North America, Western Europe and Japan.

As things stand at the moment, North America is by far the largest consumer of natural gas in the world, consuming up to 70 per cent of all gas consumed among the Western industrial nations. With major national problems of pollution, Japan is seeking to develop her natural gas supplies most of which will be in the form of imported LNG. The most reliable predictions available suggest that by the year 2000 the demand for imported natural gas will be running at around 4 million barrels of oil equivalent for Western Europe, 3 MBDOE for North America and 1.5 MBDOE for Japan, which adds up to 8.5 MBDOE total for imported natural gas of one sort or another.

The two chief clouds on the horizon for developing an international trade in natural gas are, on the one hand, political and the other, financial and commercial. The political risk is that the greater part of natural gas imports will almost certainly derive from OPEC; the commercial risk is that the LNG system is very expensive.

7.4 NUCLEAR ENERGY

The conventional wisdom of the moment is that nuclear power will substantially bridge the gap which follows in the wake of the decline in fossil fuel resources near the end of the century. It will achieve this by providing a very high proportion of the world's electricity. This belief arises from what has been assumed to be the relatively low cost of nuclear electricity, though this calculation by being made on the basis of cost at the source of production rather than of end-use can be seriously questioned. In addition the capital costs of nuclear power plants have been escalating at a phenomenal rate, as we shall examine later in this chapter. The other claim made by the proponents of nuclear electricity is that, contrary to popular opinion, it has an outstanding safety record over the last twenty-five years, especially if compared with fuels like coal, natural gas and oil. Nevertheless, because of its radioactivity and its enormous potential for destruction in its most peaceful utilisation, not to mention the possibility of the widespread proliferation of nuclear know-how and ultimately nuclear weapons, it is widely feared and strenuously opposed in many countries. There are among its critics those who are opposed to nuclear energy root and branch and those at the other extreme who merely seek to prevent nuclear power becoming the predominant source of energy in any country. The majority of those who are critical of nuclear power on environmental, economic, or political grounds stand somewhere in between these two poles.

In the environmental sphere the chief cause of contention is the means of containment of radioactivity. This fundamentally covers the processes of safe transport, storage and treatment of spent fuel elements. These elements are estimated to remain active for hundreds of thousands of years and thus constitute a legacy to posterity with which no previous generation has been confronted. Because of the multiplicity of unknowns, and because of the strong feelings known to be held by significant minorities, especially in Western countries where the possibility of influencing energy decisions, theoretically at least, remains possible, the overall future contribution of nuclear energy is still uncertain. Nonetheless some forecasting on the basis of trends and foreseeable future factors is in order.

The potential contribution of nuclear power to world's primary energy needs in 1985 and 2000, if the various nuclear programmes already existing are sustained and expanded as planned, can roughly be calculated in the following terms, namely maximum and minimum likely levels of installed nuclear electric capacity. The maximum figure is based on nuclear energy being the principal replacement for the current fossil fuels; the minimum figure is based on coal as the principal replacement for oil. The maximum likely projections would require an annual growth rate of 14 per cent per year in nuclear energy for a twenty-five year period. This compares with a minimum likely projection figure of 11 per cent per annum over a similiar period. Both these projections begin from a very low base figure of current installed nuclear capacity.

To summarise the position very briefly in percentage terms, in 1974 a mere 2 per cent of all non-Communist countries' primary energy came from nuclear power; by 1985 the maximum likely percentage will probably be around 9 per cent or a minimum of 6 per cent; by the year 2000, the maximum likely is 21 per cent and the minimum, 14 per cent. Most significantly, the maximum percentage of nuclear energy possible under the most favourable conditions is no more than 21 per cent. The important point to record is that by 2000, more than three times more than the nuclear figure would need to come from non-nuclear sources. Moreover, to put such a maximum target into even more immediate recent perspective, the maximum figure for nuclear energy in 2000 is equal to that of oil consumption in the non-Communist world in 1975.

To move on to the vital decision-making process about the various types of nuclear programmes which are much misunderstood, there are three stages of choice in nuclear programmes. First, there is a reactor operation which involves a single use of uranium without fuel reprocessing. Second, the processing of used fuel to extract and recycle plutonium and uranium and thirdly, the operation of fast breeder reactors. Most of the political debate over nuclear energy revolves around the second and third types of programme. Moreover, since the first type sometimes takes from six to ten years to install, it is crucial to lay the foundations for any future expanded programme. In practice the first type is often withheld when problems remain unresolved in the second and third, which is not strictly logical or necessary.

If you subscribe to the orthodoxy of nuclear expansion, an orthodoxy which later in this chapter we shall seriously question, beyond 2000 there are challenges which will arise which need to begin to be met within the next few years, such as the cumulative nature of nuclear energy programmes. Most notably, a high capacity nuclear expansion can only be maintained if fast breeder reactors — which both burn and produce nuclear fuels — become commercial around 1990–2000. In other words, the most controversial programme, the fast breeders, are crucial to the longer-term expansion programme. Meanwhile, substantial reserves of uranium need to be discovered and developed.

As of now fast breeder reactors are being developed by the United States, Britain, France, Japan and the Debenelux countries in a joint venture — ie between Germany, Belgium, the Netherlands and Luxembourg — also a combined venture involving electricity companies from France, Italy and Germany. Without too extended an analysis, it must be recorded that the United States programme was temporarily halted in 1977, while the British and German programmes are also being slowed by a creeping public awareness of the gravity of the commitment being made in this sphere. With a new technology, long lead times and now political protest, the scale of nuclear expansion by 1990 remains quite uncertain. The breeder may not even contribute 5 per cent of the non-Communist world's energy by 2000, so its contribution should not be exaggerated, even if it should be allowed to be developed at its maximum possible rate, which appears to be very improbable.

7.5 HYDRO-ELECTRICITY

Unlike nuclear power, hydro-electric power is already a major source of contemporary energy around the world. Moreover, it is most likely to continue to make an important contribution, especially in the less developeed countries where it is estimated less than 4 per cent of the potential sites have been developed. However, though the technical feasibility for hydro-electricity is very great in developing countries, many of these sites are far from accessible and the consequent long lead times for construction are further extended over the horizon. There is the further restraint that some of the alternative energy sources discussed later in this chapter become more competitive when questions such as accessibility, not only for building the dam but also transporting the electric power, are all taken into due account.

One of the factors which might induce the fullest exploitation of some of the more remote hydro-eclectric dam sites in the less developed countries would be if some of the developed countries calculated that it made sense to build some of the more energy intensive industrial activities, such as aluminum production, in the less developed countries. Apart from the potential abundance of hydro power they also possess a very high proportion of unexploited raw materials and a growing wish to develop local industries. While in the late-1970s the greater part of hydro-electric power consumed in the non-Communist world is produced and consumed in developed countries, the lack of further favourable sites and the growth of environmental restraints will probably drastically limit hydro expansion in the industrial countries.

7.6 MISCELLANEOUS FOSSIL FUELS

There are three principal fossil fuels besides those already discussed, namely heavy oil, oil sands and shale oil. Each of these three can be converted into liquid fuels roughly similiar in character to those products refined from crude oil. Because of this similarity of end-product, they each can be fed into the existing energy infrastructure which is among the strongest arguments commending their use in the eyes of some energy analysts. There is not much doubt that the reserves of these three fuels are very large indeed, especially compared with the fast depleting reserves of conventional oil. The current production of these three alternative fossil fuels is, however, quite restricted since the capital and operating expenses are still far greater than for conventional oil. There are in addition significant environmental problems which would need to be surmounted if production of these alternative fossil fuels were to be expanded. Above all, world oil prices would need to climb quite substantially in order to stimulate any large-scale production of these fuels.

The principal reserves of heavy oil and oil sands are believed to lie in Venezuela and Canada. Taken together, the resource base of these two fuels in the two countries amounts to about 2,000 billion barrels of oil, although less than half of this is regarded as recoverable. Meanwhile the largest known shale oil reserves exist in the United States, with lesser but still significant quantities lying in Brazil, the USSR and in China.

7.7 ALTERNATIVE FUELS

Geothermal steam energy is among those minor sources of current energy championed by some as a future major energy source. However, the prospects are not all that promising. Although natural steam is economically competitive, the resource base is very strictly limited since it demands the fairly rare geological combination of hot rocks, an underground water system and an impermeable caprock for trapping the steam and providing pressure. Many advocates of geothermal power argue for the much wider use of hot dry rocks but the technology for its commercial use is not yet sufficiently advanced to justify too much optimism at this point. However, as with all the alternative energy sources, there can be striking breakthroughs in the technological spheres, so the future contribution of geothermal energy should certainly not be discounted.

More promising at this point are the other alternative fuels such as solar heat, solar electricity, wind, tidal and the other renewable resources which will command an increasing allocation of research and development resources. Hitherto they have been very largely deprived of substantial funds because of the competing demands of other major conventional fuels, most notably by nuclear energy which has been quite the most greedy of research finance of all energy sources.

At the heart of the new appeal of the alternative energy sources, ie of the renewable variety, is the overwhelming fact that they are clean, sustainable and of almost unlimited extent. Solar water heating and space heating are already economically competitive in several countries and will become more widely so if other forms of energy become scarcer and more expensive. However, what is still true is that the storage of energy for sunless and windless periods constitutes a brake on its utilisation. Less justified is the view that investment costs are extremely high as are operating and maintenance costs.

Even those observers who assess the contribution of renewable resources as fairly minimal by the year 2000 will admit that the possibility of breakthroughs of either a technical or cost variety could drastically alter such a prospect. Moreover, the same observers are likely to acknowledge that if the economic and technical bottlenecks could be overcome, there are distinct advantages in the decentralisation of production offered by a renewable energy system which tends to have a humanising effect by its localisation of production, use and control of renewable energy. The applicability of renewable energy systems in the less developed countries is likely to greatly extend the benefits of agricultural development, which has been traditionally retarded by inefficient energy use or its absence because of the costs of transporting conventional energy from long distances. Renewable energy by its nature is local in character.

While the claims of renewable energy sources to research and development finance are argued on the basis of their critically important role beyond the year 2000, when oil and gas will be largely depleted and coal and nuclear power either limited by resource or environmental constraints, there are in fact even more pressing grounds for their encouragement and support on a greatly enlarged scale.

This reintroduces a recurring theme of this book, namely, the indispensable role that the price mechanism has to play in both producing and apportioning the balance of energy resources to the maximum advantage. When the internal energy price of the major consuming countries is allowed to climb, the possibility of conservation policies being effectively introduced and renewable energy development increased becomes a live option. The problem, as we have seen in earlier chapters, is that there is a wide variety of opinion on how best to restrain domestic energy demand on the one hand and yet at the same time to provide incentive sufficient to encourage domestic energy production of whatever sort.

There must be, on the evidence of the individual national energy policies described in this book, considerable support for the view that the price mechanism, while by no means an exclusive solution, is by far the best means yet discovered to simultaneously restrain demand and encourage domestic production. The price mechanism in turn needs reinforcement in sensible taxing policies toward both the individual consumer and potential energy producer, but price remains the principal energy policy tool so far devised in free societies.

7.8 SUMMARY PROSPECTS

The assumption underlying this final chapter is that by making certain broad assessments and estimates, the choices which will eventually have to be made by energy planners become that much clearer. Naturally, these choices will have to be made within the context of a much broader economic framework embracing both commercial and strategic criteria. Certain aspects of post-1985 energy prospects need special underlining.

(1) Energy will become an increasingly central issue in the final decades of the century. While various potential energy shortages are avoidable, the transition from a predominantly oil energy system to something else is

challenge not only to the global energy system but the world economy as a whole.

(2) The transition from the oil era to its successor, while it will be made within the context of the national and international policies and assumptions that we have described, will be largely determined by thousands of individual decisions of both public servants and private citizens who will be concerned with such issues as balance of payments, inflation, employment, etc, as with the more precise costs and benefits.

(3) The interdependence of the global energy economy will continue for a very long time to come and will at times be likely to prove critical to the proper functioning of a stable economic order. Not only are fuel resources very unevenly distributed worldwide, but a majority of nations must remain heavily dependent on OPEC. The extent to which the OPEC nations can use their position to alter radically their relative position in international trade as a whole — in the sphere of commodity prices, the transfer of capital and technology — remains to be seen. Meanwhile, among the importing countries, the United States, by virtue of the size of its requirements, its domestic energy resources and its financial capability, is likely to be of decisive importance in the course it adopts.

(4) In the future, other nations besides OPEC and the United States may become critical in their importance to the global energy economy, though by and large the latter will remain central. Thus coal from North America and some southern hemisphere countries, uranium from Australia, Canada and the United States, gas from OPEC countries, solar production of hydrogen and methanol in certain less developed countries, even an expanded export or import policy by either the Soviet Union or China, any one or several of these possibilities could prove crucial to the success or failure of future energy supply.

(5) There is an acute need for educating both the peoples of the developed and less developed countries of their mutual interdependence overall, particularly when their

interests come into conflict, which inevitably they do from time to time.

(6) There is a future need to create new and flexible forms of international co-operation in energy, much of it in the financial sphere where commercial investments need guaranteeing and where joint research programmes need help on the road to commercial utilisation. National groups within a co-operative framework may well prove a pattern worth emulating.

The foregoing analysis of global energy prospects in the post-1985 period is of necessity of a very broad nature. It is not so much a precise series of forecasts — they would be of fairly dubious reliability — but more a description of the elements in the global energy economy likely to exist during this period. However, there is a need to come to grips with some firmer energy forecasting figures if we are to do more than respond passively to whatever trends happen to be in the pipeline. The truth is that this can only be done by means of examining a national energy trend path since the global energy economy post-1985 contains far too many variables to inspire confidence in any sort of forecast.

As we have continually re-emphasised in this book, there is really only one national energy policy which dwarfs all others in its long-term influence on the global energy economy as a whole, that of the United States. In the chapter on United States energy policy we attempted to describe the strengths and weaknesses of current US energy policy and its most likely immediate consequences both domestically and internationally. Now the opportunity exists to proffer a much more radical and probing critique of long-term US energy policy. That we have included such a critique in the final chapter chiefly concerned with global trends is simply to underline the fact that what is said about long-term US trends is a pointer to the radical reappraisal required by virtually every Western industrial nation. If the United States is facing an enormous challenge to readjust already, the following analysis is to remind ourselves that the future readjustment for all Western industrial nations is likely to be no less demanding.

7.9 A RADICAL REAPPRAISAL

As we have discussed it so far, the energy policy likely to be followed over the next several decades in the United States shows signs of following a path which is essentially an extrapolation of the recent past, however much Preseident Carter may have set out to carve a new set of policies, especially in the realm of conservation. The common feature of the Nixon, Ford and Carter energy policies is that they have each rested on a steady expansion of centralised high technologies; moreover, they have all contributed to a striking trend toward increasing domestic energy supplies in general and electricity in particular.

The radical alternative to such an extrapolation of past energy policies lies in the direction of the immediate improved use of current energy and the simultaneous concentration on the development of renewable energy sources together with certain transitional fossil fuel technologies. The immediate and obvious question is why should such a radical and presumably costly new direction in energy policy be adopted? The short answer is, for the sake of the socio-political structures that are likely to arise and for sound but scarcely examined economic reasons. The more detailed answer to the question why a radical new direction should be adopted now follows. Before doing so it should be stressed that the two broad options are ultimately self-excluding since the pursuit of the present trend represents a commitment of investment capital on such a scale that the second, non-centralised option, would be effectively foreclosed. Among the most notable by-products of the second and radical alternative path would be the virtual elimination of nuclear proliferation.

Breaking it down in greater detail, of each of the Nixon, Ford and Carter administrations — ie those post-1973 when a national energy policy became a top priority — it can be said that their avowed energy objectives have included maintaining domestic supply, limiting consumption and checking the rise in imports. In pursuit of such objectives they have each sought to promote the expansion of three main sectors: firstly, of coal, chiefly by strip mining and the conversion of such coal into electricity or synthetic fluid fuels; secondly, of oil and gas increasingly from the north slope of Alaska and from continental offshore wells; thirdly, of nuclear fission of one kind or another, eventually of fast breeder reactors. The readily apparent assumption of such short to medium-term policies is that the long-term sources of US domestic energy will be a combination of solar electricity and fission and fusion breeders. Shortly after 2020, if present trends continue, nuclear power will have reached parity with oil and gas combined, with coal production roughly equal to the other three put together.

Breaking such a projection down still further, the possibility is that by the year 2000 there would be somewhere between 500 and 800 nuclear reactors, a similiar number of coal-fired power stations, 1000 to 1600 new coal mines and possibly 15 million electric cars. The electrical supply system, which may have already doubled in the decade leading up to 1985, will become the predominant form of end-use energy. If such trends are maintained, and they have already established a considerable degree of momentum, the massive electrification ensuing could prove the most important modification in the infrastructure of industrial society, certainly in North America, since the railroad.

The dangers of these trends have often been enumerated yet somehow or other have escaped any sustained criticism on economic grounds, which is surprising since they constitute the most serious objection to following a path leading to a predominantly electrified and pre-eminently nuclear economy. Today, two thirds of all fuel consumed is direct or unconverted fuel. By 2000, if present trends are maintained, electrification will consume 50 per cent of the total fuel input. The fundamental weakness of this trend towards electrification from an economic standpoint is the increasing ineffeciency, with up to *two thirds* of the original fuel being wasted. Even possible coal conversion wastes only about a third. But there is an even more disturbing trend which can be readily detected, namely a steady pattern of increasing capital intensity in electrification.

To illustrate, North Sea oilfields already onstream represent a capital investment of roughly 10,000 US dollars to deliver an extra barrel per day. In the 1980s, when Alaskan and US offshore oil becomes a significant source of supply, the estimates reach as high as 25,000 US dollars. These astronomical increases in capital costs are widely recognised within the industries concerned but because electrical capacity has normally been calculated per installed, not delivered, kilowatt, they have been disguised from the wider public. The incontrovertible truth is that the capital cost of the new systems that create electricity are far greater than for those which burn fuels directly.

To be concrete about such an assertion, for coal-electric capacity ordered in the late-1970s the potential cost might run as high as 150,000 US dollars for the delivered equivalent of one barrel of oil per day. If this seems high then it is almost trivial by comparison with nuclear-electric capacity which costs in the range of 200,000 to 300,000 US dollars per barrel of oil equivalent, calculated on a delivered capacity basis. The overall evidence suggests that the capital cost per delivered kilowatt of electric energy is generally about 100 times greater than the traditional direct fuel technologies upon which our present industrial society operates. That this most fundamental fact has been more or less concealed from the public at large has more to do with the defence of vested interests such as those of the nuclear and coal lobbies, both historically well entrenched where they are not protected, than any conspiratorial faction. The priority given to nuclear power naturally cannot be disassociated with the defence lobby which is not to say the strategic factor should be discounted in any weighing of the advantages of nuclear power, only that arguments should be deployed explicitly and openly rather than implicitly and surreptiously.

7.10 CONSERVATION PRINCIPLES

At root there are only two broad approaches to energy conservation in a modern Western industrial society: that of the technical fix, in which no radical change in life-style is demanded and that of a substantial decrease in consumption of energy, involving a marked change in life style. The first, the technical fix path, is built upon such devices as improved furnaces and cars, less extravagant lighting and ventilation in commercial buildings, thermal insulation and heat pumps. The apparently mundane nature of these changes should not disguise their enormous potential for conserving energy without too much readjustment. The second path is the more radical one and unsurprisingly demands major readjustments, though not necessarily painful ones if they are embarked upon judiciously. The radical path embraces such measures in old-fashioned thrift as car pooling, smaller cars, greatly extended mass transit systems, greater deployment of bicycles, the encouragement of walking by the provision of more footpaths and walkways, the encouragement of an 'open window' society, and a general enlargement of the practice of recycling materials.

The fact that there is enormous scope for technical fixes in the United States can be seen by comparing its energy efficiency with that of Sweden, which uses one third less energy per capita, or West Germany whose per capita energy intensity is half that of the United States in space heating and one quarter in transport. Almost the entire difference between the United States performance and that of her affluent West European counterparts is due to technical fixes. In industrial terms, among the most important means of energy conservation is in cogeneration, that is the generating of electricity as a by-product of the process steam normally produced in many industries. A recent Dow study suggested that by 1985 US industry could meet roughly half its own electricity needs — compared with around 15 per cent today — by means of cogeneration. Such a programme of cogeneration might save anything from 20 to 50 billion US dollars in investment; moreover it might also save the equivalent of 2.3 million barrels of oil per day into the bargain, not to mention reducing the price of electricity to consumers. Indeed, highlighting the situation in round terms, so great is the scope for technical fixes that it has been calculated that the United States could spend several hundred billion dollars on them for structural changes' and several hundred million more on a current basis, and still manage to save money compared with increasing the supply of energy (under any extrapolation of the present trend).

The fundamental question has therefore to be asked, what factors have prevented the self-evident need for greater energy efficiency through technical fixes taking place before now? The answer is not a single, straightforward one, but a multiplicity of answers. Yet the overall answer is neither technical nor economic obstacles but at root a myriad of institutional barriers. Some of those institutional barriers have been skilfully identified by President Carter's energy programme, whose strong suit is conservation; others have not and might be usefully listed.

These barriers embrace such perennial and almost universal features as a lack of innovation in the building industry, often because of conflicting and obsolete building codes; a lack of machinery to ease the transition from one employment to another, most especially acute when the transfer happens to shift employment from skilled to relatively unskilled workers; electricity and gas services designed to boost consumption, aided and abetted by a fee structure to building engineers that rewards increased energy consuming capacity; inappropriate tax and mortgage policies which aid higher energy consumption; and, not least, conflicting signals to energy consumers, sometimes advertising to boost consumption, sometimes advertising to restrain consumption of the same energy source. Finally, in terms of boosting the development of the soft or renewable energy sources, improved access to capital markets for the small investor in local energy structures, is vital to the progress of that sector.

Thus, to conclude with a positive response of a general kind to the challenge of overcoming the specific institutional barriers to effective technical fixes we can but return to a recurring theme of this book. Namely, if an energy path which is likely to be both efficient and humane is to be pursued, the most important single principle which must be kept paramount is to employ properly the markets that already exist. This might include such key elements as flat or even inverted rates for gas and electrical services (ie rather than discounts for large users as currently); pricing energy of whatever sort as close as possible to what extra supplies will cost in the longer term; the systematic removal of subsidies of any sort.

It might, with benefit to everyone, involve policies which embrace the natural equilibrium of the market system by doing two things: first, assessing the total costs of energy-using purchases over their whole operating lifetime, sometimes called 'life-cycle' costing; second, calculating the costs of entire energy systems, especially the support and distribution systems normally left out of the costing estimates of those promoting the expansion of their particular energy sector with consequent distortion of the ultimate real costs.

Other means of overcoming institutional barriers to energy end-use efficiency would include the proper assessment and charging of environmental costs, the valuing of assets by what it would cost to replace them, proper and consistently maintained discounting of energy infrastructure assets, all of these would tend to create a more efficient distribution of capital within the total energy production and distribution sectors.

Finally, as a means of introducing the disciplines of the market into a sector of Western industrial society which has become the victim of endemic intervention and myriad regulation, there needs to be an encouraging of competition through anti-trust enforcement. Whether this would include horizontal divestiture of the leading energy corporations needs a detailed and objective appraisal before any decision to enforce or not to enforce particular situations. None of the foregoing broad suggestions designed to lift the various forms of protection which hamper competition within the overall energy sector is intended to suggest that the market can perform every function, only that it has an indispensable and central function to play in any pluralist society.

If we are to draw any basic conclusions from this final brief summary of the means to make energy policy both more efficient and more humane, it is that the best path will combine elements of the technical fix and the change in values and life-style. Without the technical fixes with their possibilities for doubling the amount of social benefit derived from end-use energy, there is no prospect of making the transition to the renewable energy resources with their demand for a more radical change in the socio-economic systems. It is important, too, to visualise the possibilities that present themselves from combining these two emphases as quite the most flexible available.

The most damning indictment of the high technology, highly centralised, high consumption energy systems which may yet prevail on the most pessimistic expectations is that they represent the adoption of the inflexible alternative. Once embarked wholeheartedly down this path there will be very little opportunity of turning back. By contrast the path this book has urged offers the prospect of continuous adaptation to contemporary requirements as determined by consumers of many different sorts. If the thrust toward centralised energy systems is allowed to continue with its ultimate expression in a plutonium society, then the gods of energy will be gods indeed with almost an absolute capacity to determine the shape of society for decades to come, without end. More than most previous generations, the power to choose lies to hand at this moment in the energy policies each nation chooses to adopt. Our sons and daughters can only hope we choose well.

General Bibliography

(listed in chronological order as they appear as sources in the text)

Leslie Grainger, *The Role of Coal in an Integrated Energy Policy* Address to Manchester Statistical Society, (December 3, 1974).

An Appraisal of the Technical and Economic Aspects of Dungeness B Nuclear Power Stations, Central Electricity Generating Board (July 1965).

Fuel Policy, Ministry of Power, HMSO, Cmnd 3438 (November 1967).

Select Committee for Science and Technology, Session 1966–7 UK Nuclear Reactor Programme, HCP 311, HMSO (1967).

Ninth World Energy Conference (Detroit, September, 1974) Survey of World Energy Resources.

Leslie Grainger, *Coal into the Twenty-First Century*, The Robens Coal Science Lecture (London, October 7th, 1974).

P. F. Chapman and N. D. Mortimer, Research Report ERG 005 Energy Research Group, Open University, Milton Keynes.

J. Price, *Dynamic Energy Analysis and Nuclear Power*, Friends of the Earth Ltd (London, December 1974).

Nuclear Energy Balances in a World with Ceilings, International Institute for Environment and Development (December, 1974).

E. C. Williams, *Energy Policy and Planning — a General View*, Institute of Fuel Conference (Lyndhurst, May, 1973).

Douglas Evans, *The Politics of Energy* (Macmillan, 1976).

Richard M. Nixon, *Address to Congress* (April 18, 1973).

Meeting Europe's Energy Requirements, National Coal Board (1962).

Bulletin of the European Communities, Supplement 4/74 Towards a new energy policy strategy for the European community.

Robert W. Campbell, *The Economics of Soviet Oil and Gas*, (Johns Hopkin University Press, Baltimore, 1968).

John P. Hardt, *Soviet Economic Policy for the 1970s*, US Congress, (1973).

Kiyoshi Kijima, *Japanese Foreign Economic Policy*, Trade Policy Research Centre, London.

Hugh Corbet, *Trade Strategy and the Asian Pacific Region*, (Allen & Unwin, London, Toronto University Press, Toronto).

Robert Carin, *Power Industry in Communist China*, Hong Kong University, Hong Kong.

Yuan-Li Wu, *Economic Development and the Use of Energy Resources*, (Praeger, New York & London).

Jeremy Russell, *Energy as a factor in Soviet foreign policy*, Saxon House, Farnborough, Hants, England. Lexington Books, Lexington, Mass, USA 1976).

First Guidelines for a Community Energy Policy (EEC Commission, December, 1968).

The Community's relations with the Energy Producing Countries Bull ECI (1974).

Promoting the Utilization of Nuclear Energy, BULL EC2 (1974).

John Bradbeer, 'Energy in the EEC', *European Studies* (1974).

R. L. Gordon, *The Evolution of Energy Policy in Europe* (Praeger, New York, 1970).

R. Bailey, 'Britain and a Community Energy Policy', *National Westminster Bank Review* (Nov. 1973).

R. Bailey, 'Consumer Governments and the Oil Crisis', *National Westminster Bank Review* (May 1974).

G. Manners, *The Geography of Energy*, Hutchinson (1971).

The Energy Situation in the Community, EEC(76) 1392.

Kenneth W. Dam, *Oil Resources; Who Gets What How?* (University of Chicago Press, 1976).

Colin Robinson, *The Energy 'Crisis' and British Coal*, Hobart Papers 59 Institute of Economic Affairs (1974).

D. I. Mackay and G. A. Mackay, *The Political Economy of North Sea Oil* (Martin Robertson, 1976).

M. M. Sibthorp, *The North Sea: Challenge & Opportunity*, Europa.

Martin Saeter/Ian Smart (eds), *The Political Implications of North Sea Oil & Gas*, IPC Science & Technology.

Simon Evans (ed), *Energy Options in the UK*, Latimer.

J. W. House (ed), *The UK Space: Resources, Environment and the Future*, Weidenfeld & Nicolson.

Martin Ryle, *The Economics of Alternative Energy Sources*.

D. C. Ion, *Availability of Natural Resources*, Graham & Trotman.

K. A. D. Inglis, *Energy: from Surplus to Scarcity*, Applied Science.

Peter Hill & Roger Vielvoye, *Energy in Crisis*, Brandts.

Andrew L. Simon, *Energy Resources*, Pergamon.

Gerald Foley, *The Energy Question*, Penguin.

Carroll L. Wilson, *Energy, Global prospects, 1985–2000*, Report of the Workshop of Alternative Energy Strategies (WAES), McGraw-Hill, 1977.

Index